NONCOMMUTATIVE CHARACTER THEORY OF THE SYMMETRIC GROUP

NONCOMMUTATIVE CHARACTER THEORY OF THE SYMMETRIC GROUP

by

Dieter Blessenohl
Christian-Albrechts-Universität Kiel, Germany

Manfred Schocker
University of Wales Swansea, UK

Imperial College Press

Published by

Imperial College Press
57 Shelton Street
Covent Garden
London WC2H 9HE

Distributed by

World Scientific Publishing Co. Pte. Ltd.
5 Toh Tuck Link, Singapore 596224
USA office: 27 Warren Street, Suite 401-402, Hackensack, NJ 07601
UK office: 57 Shelton Street, Covent Garden, London WC2H 9HE

British Library Cataloguing-in-Publication Data
A catalogue record for this book is available from the British Library.

NONCOMMUTATIVE CHARACTER THEORY OF THE SYMMETRIC GROUP
Copyright © 2005 by Imperial College Press

All rights reserved. This book, or parts thereof, may not be reproduced in any form or by any means, electronic or mechanical, including photocopying, recording or any information storage and retrieval system now known or to be invented, without written permission from the Publisher.

For photocopying of material in this volume, please pay a copying fee through the Copyright Clearance Center, Inc., 222 Rosewood Drive, Danvers, MA 01923, USA. In this case permission to photocopy is not required from the publisher.

ISBN 1-86094-511-2

Printed in Singapore by World Scientific Printers (S) Pte Ltd

Für Conni und Ingke

Preface

This set of lecture notes presents a new approach to the representation theory of the symmetric group — more precisely: to the character theory of the symmetric group over a field of characteristic zero. Knowledge of the classical theory is unnecessary — perhaps, even a hindrance to the understanding of this new theory which is in many ways more immediate, more efficient, more transparent, and more elementary due to the new tools that lie behind the word "noncommutative".

The ordinary character theory of the symmetric group is a commutative theory based on the commutative algebra of class functions or, equivalently, the algebra of symmetric functions. The monographs of James and Kerber [JK81] and of Macdonald [Mac95] are general references on the subject.

The noncommutative character theory is based on a noncommutative algebra which maps onto the algebra of class functions. The underlying idea is to transfer problems to the noncommutative superstructure, and to solve them in that setting.

Louis Solomon was the first to introduce this idea onto the character theory of the symmetric group, in a seminal article of 1976. Since then, Armin Jöllenbeck and Christophe Reutenauer have been among the main architects of the theory, as well as the members of the research group in Paris: Gérard Duchamp, Israel Gelfand, Florent Hivert, Daniel Krob, Alain Lascoux, Bernard Leclerc, Jean-Yves Thibon, to name a few. Our presentation of the material grew out of several lecture courses held in Kiel during the past years. It is self-contained and suitable for undergraduate level.

The main text is subdivided into three parts. Basic properties of the algebra of class functions are rederived in Part I. Part II contains the construction of the noncommutative superstructure and plays the key role here.

Various applications are then given in Part III.

We have included three appendices to supplement the main text, in rather different ways. Appendix A contains a short and self-contained dispatch of the character theory of finite groups in general, to the extent needed here. Appendix B contains a simple proof of Solomon's theorem. In Appendix C, a new approach is presented, to the Robinson-Schensted correspondence and related results of Knuth and Schützenberger which are at the combinatorial core of noncommutative character theory.

The reader is referred to the introduction for much more details on the ideas of the noncommutative theory and the thread of this manuscript.

We wish to express our thanks to the Research Chairs of Canada and, in particular, to Deutsche Forschungsgemeinschaft for generous support through the last years*. We would also like to thank Adalbert Kerber. This project would not have been possible without him. Besides, sincere acknowledgement is due to those who helped us by a careful scrutiny of the manuscript: *thank you*, Bill Boshuk; *þakk*, Wilbur Jonsson; *merci beaucoup*, Luc Lapointe; *vielen Dank*, Christophe Reutenauer.

September 2004

*Research projects DFG BL-488 and DFG Scho-799.

Contents

Chapter 1

Introduction

Representation theory is the study of groups G and algebras A by means of homomorphisms

$$d : G \to \mathsf{GL}_K(V) \quad \text{and} \quad D : A \to \mathsf{End}_K(V),$$

into the group of invertible linear endomorphisms, respectively, the algebra of endomorphisms of a vector space V over a field K. This brings powerful tools of linear algebra to bear on the theory of groups and algebras. The group algebra KG of G over the field K connects the two theories, since, by linearity, any *representation* $d : G \to \mathsf{GL}_K(V)$ of G extends to a *representation* $D : KG \to \mathsf{End}_K(V)$ of KG, which is *unital* in the sense that the image of the identity element of G under D is the identity of $\mathsf{End}_K(V)$. By restriction, any unital representation $D : KG \to \mathsf{End}_K(V)$ defines a homomorphism $d : G \to \mathsf{GL}_K(V)$. The representation theory of groups may thus be viewed as a special case of the representation theory of algebras with identity, with the restriction to unital representations.

The theory of *modules* for groups and algebras works equally well, because any unital representation d of G, or D of A, turns V into a unital G-module, or A-module, by setting

$$vg := v(gd) \qquad \text{or} \qquad va := v(aD),$$

for all $v \in V$ and $g \in G$, $a \in A$. Conversely, any unital G-module, or A-module, provides, in an obvious way, a representation of G, or A. In what follows, all modules are assumed to be unital and finite-dimensional, since the representation theory of finite symmetric groups is the focus of this study.

A linear subspace N of an A-module M is an A-*submodule* of M if N is closed under the action of A, that is, if $na \in N$ for all $n \in N$ and $a \in A$.

The building blocks of the representation theory of an associative algebra A with identity are the *simple* or *irreducible A-modules* $M \neq \{0_M\}$ whose only A-submodules are $N = \{0_M\}$ and $N = M$. An A-module M is *semi-simple* or *completely reducible* if M is the direct sum of irreducible A-modules. Equivalently, any A-submodule N of M has an A-module complement in M, that is, there exists an A-submodule N' of M such that $N \cap N' = \{0_M\}$ and $M = N + N'$.

Assuming right multiplication, the algebra A is an A-module which is called the *regular A-module* and denoted by A^R. The A-submodules of A^R are the right ideals of A. The algebra A is called semi-simple if the regular A-module A^R is semi-simple.

It can be shown that, in this sense, if A is semi-simple, then so is every A-module.

The principal tasks of the representation theory of semi-simple algebras may be summarised as follows:

1. Classify the isomorphism classes of irreducible A-modules and display a representative of each class.
2. Find ways to decompose an arbitrary A-module into irreducible A-submodules.

In the representation theory of finite groups, the following result is crucial.

Maschke's Theorem. *If G is a finite group and K is a field of characteristic not dividing the order of G, then KG is semi-simple.*

Let M be a G-module and let $d : G \to \mathsf{GL}_K(M)$ denote the corresponding representation of G. The mapping

$$\chi_M : G \to K\,, \quad g \mapsto \mathsf{tr}(gd)$$

is the *character of G afforded by M*, where $\mathsf{tr} : \mathsf{End}_K(M) \to K$ denotes the trace function. It is readily seen that χ_M is constant on the conjugacy classes of G. Any such map $G \to K$ is a *class function* of G. The linear space of all class functions of G is denoted by $\mathcal{C}\ell_K(G)$.

Let KG be semi-simple, then the character afforded by M indeed "characterises" M. For, in this situation, two G-modules M and M' are isomorphic if and only if $\chi_M = \chi_{M'}$. We say that χ_M is irreducible if M is irreducible. Then any character of G is a linear combination of irreducible characters with nonnegative integer coefficients, and these coefficients determine the underlying G-module, up to isomorphism. If, in addition, K is a so-called *splitting field* of G, there is the explicit formula

$$\alpha = \sum_{\chi} (\alpha, \chi)_G \, \chi$$

expressing any $\alpha \in \mathcal{C}\ell_K(G)$ as a linear combination of the irreducible characters χ of G, with coefficients given by the scalar product

$$(\alpha, \chi)_G = 1/|G| \sum_{g \in G} \alpha(g^{-1})\chi(g).$$

As a consequence, for any G-module M, there is the isomorphism of G-modules

$$M \cong_G \bigoplus_{\chi} (\chi_M, \chi)_G \, I_\chi ,$$

where I_χ is an irreducible G-module affording χ, for each irreducible character χ of G. A short and self-contained dispatch of the theory of finite group characters, up to this point, is contained in Appendix A.

If the characteristic of K is positive and divides the order of G, then the group ring KG is not semi-simple. The so-called *modular* representation theory of G arising therefrom is of a totally different nature. Therefore, it is assumed throughout that K *is a field containing the field* $\mathbb{Q}$ *of rational numbers as a subfield.*

* * *

Let $\mathbb{N} := \{1, 2, 3, \ldots\}$ be the set of positive integers. Put $\mathbb{N}_0 := \mathbb{N} \cup \{0\}$ and

$$\underline{n} := \{1, \ldots, n\}$$

for all $n \in \mathbb{N}_0$. The symmetric group $\mathcal{S}_n$ consists of all bijections (or permutations) $\pi : \underline{n} \to \underline{n}$. In examples, we write $\pi \in \mathcal{S}_n$ as a word or as a product of cycles as usual. For instance, the permutation

$$\pi := \begin{pmatrix} 1\,2\,3\,4\,5\,6\,7\,8\,9 \\ 4\,1\,3\,6\,5\,2\,7\,9\,8 \end{pmatrix} \in \mathcal{S}_9$$

reads as $\pi = (1\,4\,6\,2)\,(3)\,(5)\,(7)\,(8\,9) = (1\,4\,6\,2)\,(8\,9)$ as a product of cycles and as $\pi = 4\,1\,3\,6\,5\,2\,7\,9\,8$ as a word. Products $\pi\sigma$ of permutations are to be read from left to right: first π, then σ. We write id_n for the identity in $\mathcal{S}_n$.

We shall see later on that $\mathbb{Q}$ is a splitting field of $\mathcal{S}_n$. In addressing the representation theory of $\mathcal{S}_n$, the general theory sketched above therefore shows that the following two problems require solutions:

1. Determine the irreducible characters of $\mathcal{S}_n$.
2. Find ways to evaluate scalar products $(\alpha, \chi)_{\mathcal{S}_n}$, where α is an arbitrary character of $\mathcal{S}_n$ and χ is an irreducible character of $\mathcal{S}_n$.

Specht modules and matrix representations of the symmetric group will not be considered here.

The symmetric group $\mathcal{S}_k$ and also the direct product $\mathcal{S}_k \times \mathcal{S}_{n-k}$ occur as subgroups of $\mathcal{S}_n$ in a multitude of ways whenever $1 \leq k \leq n-1$. In this sense, any character of $\mathcal{S}_n$ yields a character of $\mathcal{S}_k$, or of $\mathcal{S}_k \times \mathcal{S}_{n-k}$, by restriction. Conversely, any character of $\mathcal{S}_k$, or of $\mathcal{S}_k \times \mathcal{S}_{n-k}$ yields a character of $\mathcal{S}_n$, by induction. There is the general idea to deduce results for $\mathcal{S}_n$-characters from the character theory of $\mathcal{S}_k$, $\mathcal{S}_{n-k}$, and $\mathcal{S}_k \times \mathcal{S}_{n-k}$, which are assumed to be "well understood", since k and $n-k$ are both $< n$. The major tools of this *inductive method* are induction and restriction of characters together with Frobenius' reciprocity law.

The inductive method may be described elegantly in terms of the *bialgebra of class functions* [Gei77], which is defined as follows. The underlying vector space of this bialgebra is the direct sum

$$\mathcal{C} = \bigoplus_{n \geq 0} \mathcal{C}\ell_K(\mathcal{S}_n).$$

To each class function α of $\mathcal{S}_k$ and each class function β of $\mathcal{S}_{n-k}$ may be associated a class function $\alpha \# \beta$ of $\mathcal{S}_k \times \mathcal{S}_{n-k}$, in a natural way (see 3.2). The product on $\mathcal{C}$ arises from the concept of induction:

$$\alpha \bullet \beta = (\alpha \# \beta)^{\mathcal{S}_n}.$$

The restriction of class functions leads to a *coproduct* on $\mathcal{C}$, that is, a linear mapping $\mathcal{C} \to \mathcal{C} \otimes \mathcal{C}$. It is defined by

$$\alpha{\downarrow} = \sum_{k=0}^{n} (\alpha|_{\mathcal{S}_k \times \mathcal{S}_{n-k}})\, i_{k,n-k}^{-1}$$

for all class functions α of $\mathcal{S}_n$. Here $i_{k,n-k}$ denotes the natural linear isomorphism $\mathcal{C}\ell_K(\mathcal{S}_k) \otimes \mathcal{C}\ell_K(\mathcal{S}_{n-k}) \to \mathcal{C}\ell_K(\mathcal{S}_k \times \mathcal{S}_{n-k})$. The coproduct $\downarrow$ is an algebra map $(\mathcal{C}, \bullet) \to (\mathcal{C} \otimes \mathcal{C}, \bullet_\otimes)$, where $\bullet_\otimes$ is the product on $\mathcal{C} \otimes \mathcal{C}$ arising from $\bullet$ (see 2.8). In other words, $(\mathcal{C}, \bullet, \downarrow)$ is a *bialgebra*. By orthogonality, the scalar products $(\cdot,\cdot)_{\mathcal{S}_n}$ on the linear spaces $\mathcal{C}\ell_K(\mathcal{S}_n)$ extend to a single bilinear form on $\mathcal{C}$. The bialgebra $\mathcal{C}$ is *self-dual* with respect to this form, that is,

$$(\alpha \bullet \beta, \gamma)_{\mathcal{C}} = (\alpha \otimes \beta, \gamma\!\downarrow)_{\mathcal{C}\otimes\mathcal{C}}$$

for all $\alpha, \beta, \gamma \in \mathcal{C}$, where $(\,\cdot\,,\,\cdot\,)_{\mathcal{C}\otimes\mathcal{C}}$ is the bilinear form on $\mathcal{C} \otimes \mathcal{C}$ such that

$$(\alpha_1 \otimes \alpha_2, \beta_1 \otimes \beta_2)_{\mathcal{C}\otimes\mathcal{C}} = (\alpha_1, \beta_1)_{\mathcal{C}}(\alpha_2, \beta_2)_{\mathcal{C}}$$

for all $\alpha_1, \alpha_2, \beta_1, \beta_2 \in \mathcal{C}$. It is this self-duality that mirrors Frobenius' reciprocity law.

The bialgebra of class functions contains the classical representation theory of the symmetric group — in terms of ordinary, that is to say, *commutative character theory.*

For example, if χ is an irreducible character of $\mathcal{S}_k$, and ψ is an irreducible character of $\mathcal{S}_{n-k}$, then $\chi \bullet \psi$ is a character of $\mathcal{S}_n$. It is a classical and fairly intricate problem to decompose this induced character into irreducible characters of $\mathcal{S}_n$. An answer is provided by the Littlewood–Richardson Rule [LR34]. In terms of the bialgebra $\mathcal{C}$, this remarkable result becomes a description of the structure constants with respect to that linear basis which consists of all irreducible characters of all symmetric groups $\mathcal{S}_n$.

Chapter 2 contains parts from the theory of coproducts and bialgebras, to the extent needed, while more details concerning the definition and basic properties of the bialgebra $\mathcal{C}$ are given in Chapter 3.

Each of the bialgebras $\mathcal{A}$ considered here is an inner direct sum

$$\mathcal{A} = \bigoplus_{n\in\mathbb{N}_0} \mathcal{A}_n$$

of linear subspaces $\mathcal{A}_n$ such that

$$\mathcal{A}_n \star \mathcal{A}_m \subseteq \mathcal{A}_{n+m} \quad\text{and}\quad \mathcal{A}_n\delta \subseteq \sum_{k=0}^{n} \mathcal{A}_k \otimes \mathcal{A}_{n-k}\,,$$

for all $n, m \in \mathbb{N}_0$ and $\mathcal{A}_0 \cong K$, where $\star$ denotes the product and δ denotes the coproduct on $\mathcal{A}$. This means that $\mathcal{A}$ is *graded* and *connected.* Any such bialgebra $\mathcal{A}$ is actually a *Hopf algebra.*

The algebra of symmetric functions Λ is isomorphic to $\mathcal{C}$. Andreij Zelevinski's approach [Zel81b] to the representation theory of finite classical groups builds on this particular type of commutative Hopf algebra.

However, knowledge of the theory of symmetric functions, or the theory of Hopf algebras, is not necessary here.

* * *

The origin of noncommutative character theory is Solomon's discovery [Sol76] of a subalgebra of the group algebra of an arbitrary finite Coxeter group W which maps into the algebra of class functions of W. His results in case $W = \mathcal{S}_n$ are briefly revisited below. In our approach, they serve as backdrop and source of motivation, and not as structural building blocks.

Some notations are needed. Many interesting objects in the theory are indexed by *partitions* or *compositions*. It is convenient to represent these indices in terms of words in a free monoid $\mathbb{N}^*$ over the alphabet $\mathbb{N}$. The multiplication in $\mathbb{N}^*$ is *concatenation*. We write $q.r$ for the concatenation product of $q, r \in \mathbb{N}^*$ in order to avoid confusion with the ordinary product in $\mathbb{N}$. Any $q \in \mathbb{N}^*$ may be written uniquely as a product

$$q = q_1.q_2.\dots.q_k,$$

where $q_1, q_2, \dots, q_k \in \mathbb{N}$. The identity element $\varnothing$ of $\mathbb{N}^*$ is the empty product. If $q_1 + q_2 + \dots + q_k = n \in \mathbb{N}_0$, then q is a *composition* of n, denoted by $q \models n$. If, in addition, $q_1 \geq q_2 \geq \dots \geq q_k$, then q is a *partition* of n and we write $q \vdash n$.

Let $n \in \mathbb{N}_0$ and $q = q_1.\dots.q_k \models n$. Denote by $P^q = (P^q_1, \dots, P^q_k)$ the set partition of $\underline{n}$ consisting of the successive blocks of order q_i in $\underline{n}$, for all $i \in \underline{k}$. For example, $P^{2.1.2} = (\{1,2\}, \{3\}, \{4,5\})$. The *Young subgroup of type* q in $\mathcal{S}_n$ is

$$\mathcal{S}_q := \{\, \pi \in \mathcal{S}_n \,|\, P^q_i \pi = P^q_i \text{ for all } i \in \underline{k} \,\}.$$

If $\pi, \sigma \in \mathcal{S}_n$, then clearly $\mathcal{S}_q \pi = \mathcal{S}_q \sigma$ if and only if $P^q_i \pi = P^q_i \sigma$ for all $i \in \underline{k}$. Therefore, the set

$$\mathcal{S}^q := \{\, \nu \in \mathcal{S}_n \,|\, \nu|_{P^q_i} \text{ is increasing for all } i \in \underline{k} \,\}$$

is a transversal of the right cosets of $\mathcal{S}_q$ in $\mathcal{S}_n$. The symmetric group $\mathcal{S}_n$ acts on these right cosets, by right multiplication. The corresponding character $\xi^q = (1_{\mathcal{S}_q})^{\mathcal{S}_n}$ is the *Young character of type* q. The set of Young characters $\{\, \xi^p \,|\, p \vdash n \,\}$ is a linear basis of $\mathcal{C}\ell_K(\mathcal{S}_n)$ (see 12.3). Observe that the linear space $\mathcal{C}\ell_K(\mathcal{S}_n)$ is a ring with multiplication defined by $(\alpha\beta)(\pi) = \alpha(\pi)\beta(\pi)$ for all $\alpha, \beta \in \mathcal{C}\ell_K(\mathcal{S}_n)$ and $\pi \in \mathcal{S}_n$.

The Mackey formula [CR62, (44.3)] yields a multiplication rule

$$\xi^r \xi^q = \sum_{s \models n} m_q^r(s)\, \xi^s$$

and a combinatorial description of the coefficients $m_q^r(s) \in \mathbb{N}_0$. Solomon's far-reaching discovery was a noncommutative refinement of this rule for the elements $\Xi^q := \sum_{\nu \in S^q} \nu$, $q \models n$, of the group algebra $K\mathcal{S}_n$:

$$\Xi^r \Xi^q = \sum_{s \models n} m_q^r(s)\, \Xi^s, \tag{1.1}$$

with the same coefficients as above. As a consequence, there is the following result.

1.1 Theorem. (Solomon, 1976) *Let $n \in \mathbb{N}$. The linear span $\mathcal{D}_n$ of the elements Ξ^q, $q \models n$, is a subalgebra of the group algebra $K\mathcal{S}_n$.*

Furthermore, the linear map $c_n : \mathcal{D}_n \to \mathcal{C}\ell_K(\mathcal{S}_n)$, defined by $\Xi^q \mapsto \xi^q$ for all $q \models n$, is an epimorphism of algebras.

Several different proofs of this result have since been given (see, for example, Tits' appendix to Solomon's original paper [Sol76], or [BBHT92; BL93; Bro00; GR89; Ges84; Reu93; vW98]). A short and transparent proof of Solomon's theorem may also be found in Appendix B.

The elements Ξ^q, $q \models n$, actually form a linear basis of $\mathcal{D}_n$. To see this, observe that for any $\pi \in \mathcal{S}_n$, we have $\pi \in S^q$ if and only if the *descent set*

$$\mathsf{Des}(\pi) = \{\, i \in \underline{n-1}_{|} \mid i\pi > (i+1)\pi \,\}$$

of π is contained in the set $\{q_1, q_1 + q_2, \ldots, q_1 + \ldots + q_{l-1}\}$ of partial sums of q. The elements

$$\Delta^D := \sum_{\mathsf{Des}(\pi)=D} \pi \qquad (D \subseteq \underline{n-1}_{|})$$

of $K\mathcal{S}_n$ are clearly linearly independent, and

$$\Xi^q = \sum_{D \subseteq \{q_1, q_1+q_2, \ldots\}} \Delta^D$$

for all $q \models n$. An inclusion/exclusion argument implies that both sets $\{\,\Xi^q \mid q \models n\,\}$ and $\{\,\Delta^D \mid D \subseteq \underline{n-1}_{|}\}$ are linear bases of $\mathcal{D}_n$. Accordingly, the algebra $\mathcal{D}_n$ is referred to as the *Solomon descent algebra* of $\mathcal{S}_n$.

The direct sum

$$\mathcal{D} = \bigoplus_{n \in \mathbb{N}_0} \mathcal{D}_n ,$$

has K-basis $\{ \Xi^q \mid q \in \mathbb{N}^* \}$. We define the structure of a bialgebra on $\mathcal{D}$, as follows. The multiplication is given by

$$\Xi^r \star \Xi^q := \Xi^{r.q}$$

for all $q, r \in \mathbb{N}^*$, and linearity, so that $(\mathcal{D}, \star)$ is a free associative algebra over the set of (noncommuting) variables $\{ \Xi^n \mid n \in \mathbb{N} \}$. As a consequence, there is a unique coproduct $\downarrow$ on $\mathcal{D}$ such that $(\mathcal{D}, \star, \downarrow)$ is a bialgebra and

$$\Xi^n \downarrow = \sum_{k=0}^{n} \Xi^k \otimes \Xi^{n-k}$$

for all $n \in \mathbb{N}$. Here, by definition, Ξ^0 is the identity $\Xi^{\varnothing}$ of $(\mathcal{D}, \star)$.

The bialgebra $(\mathcal{D}, \star, \downarrow)$ is isomorphic to the *algebra of noncommutative symmetric functions.* The latter was introduced by Gelfand et al. in [GKL+95] and has been further studied extensively in so far five subsequent papers [KLT97; DKK97; KT97; KT99; DHT02].

Defining a bilinear form on $\mathcal{D}$ by

$$(\Xi^r, \Xi^q)_{\mathcal{D}} := |\mathcal{S}^r \cap (\mathcal{S}^q)^{-1}|$$

for all $q, r \in \mathbb{N}^*$, we are in a position to state and proof the following result.

1.2 Theorem. *The linear map $\mathcal{D} \to \mathcal{C}$, defined by $\Xi^q \mapsto \xi^q$ for all $q \in \mathbb{N}^*$, is an isometric epimorphism of bialgebras with respect to $(\cdot,\cdot)_{\mathcal{D}}$ and $(\cdot,\cdot)_{\mathcal{C}}$.*

The homomorphism rules for the products and the coproducts on $\mathcal{D}$ and $\mathcal{C}$ follow from the definition of $(\mathcal{D}, \star, \downarrow)$ and the corresponding rather immediate identities in $\mathcal{C}$:

$$\xi^r \bullet \xi^q = \xi^{r.q} \quad \text{and} \quad \xi^n \downarrow = \sum_{k=0}^{n} \xi^k \otimes \xi^{n-k}$$

for all $r, q \in \mathbb{N}^*$ and $n \in \mathbb{N}$ (see 3.2 and 3.9).

The interesting fact that the simultaneous linear extension of Solomon's epimorphisms c_n is an isometry, follows from Solomon's theorem: observe first that the identity $\mathcal{S}^n = \{\mathrm{id}_n\}$ and Frobenius' reciprocity law imply

$$(\Xi^q, \Xi^n)_{\mathcal{D}} = 1 = (1_{\mathcal{S}_q}, 1_{\mathcal{S}_q})_{\mathcal{S}_q} = ((1_{\mathcal{S}_q})^{\mathcal{S}_n}, \xi^n)_{\mathcal{S}_n} = (\xi^q, \xi^n)_{\mathcal{C}}$$

for all $q \models n$. Now, for arbitrary $q, r \models n$, comparing the coefficient of id_n on both sides of (1.1), gives

$$\begin{aligned}(\Xi^r, \Xi^q)_{\mathcal{D}} &= \sum_{s \models n} m_q^r(s)(\Xi^s, \Xi^n)_{\mathcal{D}} \\ &= \sum_{s \models n} m_q^r(s)(\xi^s, \xi^n)_{\mathcal{C}} \\ &= (\xi^r \xi^q, \xi^n)_{\mathcal{C}} \\ &= (\xi^r, \xi^q)_{\mathcal{C}}.\end{aligned}$$

As a consequence of Theorem 1.2, there is a noncommutative character theory which is satisfying at least to some extent. Every $\mathcal{S}_n$-character has a counterpart in $\mathcal{D}_n$ and the epimorphism $\mathcal{D} \to \mathcal{C}$ allows one to transfer every problem involving restriction, induction and scalar products of characters. However, this noncommutative setup builds on the formal procedure of passing from a polynomial ring in the set $\{\xi^n \mid n \in \mathbb{N}\}$ of commuting variables to the free associative algebra over the set $\{\Xi^n \mid n \in \mathbb{N}\}$ of noncommuting variables. This turns out to be too restrictive, even in view of the classical results on irreducible characters of $\mathcal{S}_n$. For, unfortunately, suitable noncommutative counterparts in $\mathcal{D}$ of irreducible characters (which should allow significantly simplified arguments in the noncommutative superstructure) cannot be found.

* * *

Due to Jöllenbeck ([Jöl98], see also [Jöl99]), there is the crucial idea to construct a proper extension of the bialgebra $\mathcal{D}$ (and Theorem 1.2), in order to fill this gap. The *bialgebra of permutations*, introduced by Malvenuto and Reutenauer in [MR95], provides the general framework for this purpose. The underlying vector space of this bialgebra is the direct sum

$$\mathcal{P} = \bigoplus_{n \in \mathbb{N}_0} K\mathcal{S}_n$$

and has the set of all permutations as a linear basis. The product in $\mathcal{P}$ may be described by

$$\pi \star \sigma := \sum_{\gamma} \gamma$$

for all permutations $\pi \in \mathcal{S}_n$, $\sigma \in \mathcal{S}_m$, where the sum is taken over all permutations $\gamma \in \mathcal{S}_{n+m}$ such that $i\gamma < j\gamma$ if and only if $i\pi < j\pi$ for all $i, j \in \underline{n}$, and $i\gamma < j\gamma$ if and only if $(i-n)\sigma < (j-n)\sigma$ for all $i, j \in \underline{n+m}\backslash\underline{n}$. For example,

$$12 \star 21 = 1243 + 1342 + 1432 + 2341 + 2431 + 3421.$$

The unique element $\emptyset$ of $\mathcal{S}_0$, the empty permutation, is the identity of $(\mathcal{P}, \star)$.

A combinatorial description of the coproduct on $\mathcal{P}$ is

$$\pi\downarrow := \sum_{k=0}^{n} \alpha_k \otimes \beta_k$$

for all $\pi \in \mathcal{S}_n$, where, for each $k \in \underline{n} \cup \{0\}$, $\alpha_k \in \mathcal{S}_k$ and $\beta_k \in \mathcal{S}_{n-k}$ are determined by the conditions that $i\alpha_k^{-1} < j\alpha_k^{-1}$ if and only if $i\pi^{-1} < j\pi^{-1}$, for all $i, j \in \underline{k}$, and $(i-k)\beta_k^{-1} < (j-k)\beta_k^{-1}$ if and only if $i\pi^{-1} < j\pi^{-1}$, for all $i, j \in \underline{n}\backslash\underline{k}$. For example,

$$4132\downarrow = \emptyset \otimes 4132 + 1 \otimes 321 + 12 \otimes 21 + 132 \otimes 1 + 4132 \otimes \emptyset.$$

A bilinear form on $\mathcal{P}$ is defined by

$$(\pi, \sigma)_{\mathcal{P}} := \begin{cases} 1 & \text{if } \pi = \sigma^{-1}, \\ 0 & \text{otherwise,} \end{cases}$$

for all permutations π and σ. Malvenuto and Reutenauer [MR95] showed that $\mathcal{P}$ is a self-dual bialgebra, and that $(\mathcal{D}, \star, \downarrow)$ as defined above is a subbialgebra of $\mathcal{P}$. The bialgebra of permutations is introduced in Chapter 5. In view of applications, it is suitable to use an approach which builds on Stanley's theory of P-partitions [Sta72], which is briefly revisited in Chapter 4 for that reason.

The algebra $\mathcal{P}$ maps onto $\mathcal{C}$, as follows. The *cycle type* of a permutation π in $\mathcal{S}_n$ is the partition p of n obtained by concatenating the lengths of the cycles occurring in the cycle decomposition of π, in a non-increasing fashion. It is well-known that two permutations π, σ in $\mathcal{S}_n$ are conjugate in $\mathcal{S}_n$ if and only if they have the same cycle type.

Let $\Pi_n \in K\mathcal{S}_n$ be primitive for all $n \in \mathbb{N}$, that is, $\Pi_n\downarrow = \Pi_n \otimes \emptyset + \emptyset \otimes \Pi_n$. Assign to $\varphi \in K\mathcal{S}_n$ the class function $c_\Pi(\varphi)$ of $\mathcal{S}_n$ which maps each π in $\mathcal{S}_n$ of cycle type $p = p_1.\dots.p_l \vdash n$ to

$$c_\Pi(\varphi)(\pi) := (\varphi, \Pi_{p_1} \star \cdots \star \Pi_{p_l})_{\mathcal{P}}.$$

This defines a linear map

$$c_\Pi : \mathcal{P} \to \mathcal{C}.$$

It is an important observation that c_Π is in fact a homomorphism of algebras from $(\mathcal{P}, \star)$ into $(\mathcal{C}, \bullet)$; see Chapter 7. Assuming a mild condition on the primitive elements Π_n (namely that the coefficient $(\Pi_n, \mathsf{id}_n)_\mathcal{P}$ of the identity in Π_n is 1), c is onto and coincides with Solomon's epimorphism c_n from $\mathcal{D}_n$ onto $\mathcal{C}\ell_K(\mathcal{S}_n)$ when restricted to $\mathcal{D}_n$, for all $n \in \mathbb{N}_0$.

The algebra map $c_\Pi : \mathcal{P} \to \mathcal{C}$ is not a bialgebra map nor an isometry. However, for properly chosen primitive elements Π_n, c_Π has these properties when restricted to the *coplactic algebra* $\mathcal{Q} \subseteq \mathcal{P}$, which contains $\mathcal{D}$ as well as suitable noncommutative irreducible characters. This bialgebra was discovered by Poirier and Reutenauer [PR95]. Its definition relies on the famous Robinson–Schensted correspondence [Rob38; Sch61]. In his thesis [Jöl98], Jöllenbeck considered a smaller extension $\mathcal{F}$ of $\mathcal{D}$ which is revisited in Chapter 6.

Standard Young tableaux of shape p are realised as permutations in what follows. The link to the usual notion of a tableau (see, for example, [Ful97]) is given by the concept of juxtaposing the rows of a Young diagram. For example, consider the usual picture

$$\begin{array}{|c|c|c|}\hline 5 & 7 \\ \cline{1-2} 2 & 6 & 8 \\ \hline 1 & 3 & 4 \\ \hline \end{array}$$

of an array of numbers increasing in rows and decreasing in columns. Juxtaposing the rows from top to bottom, yields the standard Young tableau

$$5\,7\,2\,6\,8\,1\,3\,4 \in \mathcal{S}_8$$

of shape 3.3.2. The set of all standard Young tableaux of shape $p \vdash n$ is denoted by $\mathsf{SYT}^p \subseteq \mathcal{S}_n$.

The Robinson–Schensted correspondence yields a bijection

$$\mathcal{S}_n \longrightarrow \bigcup_{p \vdash n} \mathsf{SYT}^p \times \mathsf{SYT}^p, \quad \pi \longmapsto (P(\pi), Q(\pi)). \tag{1.2}$$

Following Schensted, its first component $P(\pi)$ is called the *P-symbol* of π, while its second component $Q(\pi)$ is called the *Q-symbol* of π. Collecting together all permutations $\pi \in \mathcal{S}_n$ with a given Q-symbol σ, we obtain a *coplactic class*

$$A_\sigma = \{\, \pi \in \mathcal{S}_n \,|\, Q(\pi) = \sigma \,\}$$

in $\mathcal{S}_n$, for all $\sigma \in \bigcup_{p \vdash n} \mathsf{SYT}^p$. We write ΣA for the sum of A in $K\mathcal{S}_n$, for all subsets $A \subseteq \mathcal{S}_n$. The coplactic algebra $\mathcal{Q}$ is defined as the linear span of the sums ΣA_σ of all coplactic classes. As already mentioned, it is a sub-bialgebra of the bialgebra $\mathcal{P}$ of permutations and contains $\mathcal{D}$; see Chapter 8.

A theorem of Schützenberger [Sch63] states that $P(\pi) = Q(\pi^{-1})$ for all $\pi \in \mathcal{S}_n$, which implies that

$$(\Sigma A_\sigma, \Sigma A_\nu)_{\mathcal{P}} = \#\{\, \pi \in \mathcal{S}_n \,|\, Q(\pi) = \sigma,\ P(\pi) = \nu \,\} = \begin{cases} 1 & \text{if } p = q, \\ 0 & \text{otherwise,} \end{cases}$$

for all $\sigma \in \mathsf{SYT}^p$ and $\nu \in \mathsf{SYT}^q$, since (1.2) is a bijection. Furthermore, for each $p \vdash n$, there exists a standard Young tableau σ of shape p such that $\mathsf{SYT}^p = A_\sigma$; hence, denoting by Z^p the sum of SYT^p in $K\mathcal{S}_n$, there are the *noncommutative orthogonality relations*

$$(\mathrm{Z}^p, \mathrm{Z}^q)_{\mathcal{P}} = \begin{cases} 1 & \text{if } p = q, \\ 0 & \text{otherwise,} \end{cases} \tag{1.3}$$

for all partitions p and q of n. Note that, by definition, $(\mathrm{Z}^p, \mathrm{Z}^q)_{\mathcal{P}}$ is equal to the number of standard Young tableaux π of shape p such that π^{-1} is a standard Young tableau of shape q. The reader is encouraged to verify (1.3) for small values of n. Appendix C contains a new proof of the Robinson–Schensted correspondence and related results of Knuth, Schensted and Schützenberger, which builds on [BJ99].

When choosing suitable primitive elements Π_n in $\mathcal{Q}_n := \mathcal{Q} \cap K\mathcal{S}_n$ for all $n \in \mathbb{N}$, we get the following concluding result of Part II, in Chapter 9.

1.3 Main Theorem. $c_\Pi|_{\mathcal{Q}} : (\mathcal{Q}, \star, \downarrow) \to (\mathcal{C}, \bullet, \downarrow)$ *is a graded and isometric epimorphism of bialgebras, that is,* $c_\Pi(\mathcal{Q}_n) = \mathcal{C}\ell_K(\mathcal{S}_n)$ *and*

$$(\alpha, \beta)_{\mathcal{P}} = (c_\Pi(\alpha), c_\Pi(\beta))_{\mathcal{C}},$$

$$c_\Pi(\alpha \star \beta) = c_\Pi(\alpha) \bullet c_\Pi(\beta),$$

$$(c_\Pi \otimes c_\Pi)(\alpha\!\downarrow) = c_\Pi(\alpha)\!\downarrow$$

for all $n \in \mathbb{N}_0$ *and* $\alpha, \beta \in \mathcal{Q}$.

This is a slight extension of a result of Jöllenbeck, who considered a particular series $(\omega_n)_{n\in\mathbb{N}}$ of primitive elements (see 9.4). The epimorphism $\mathcal{P} \to \mathcal{C}$ associated with this series will simply be denoted by c. Any inverse image under c in $\mathcal{P}$ of a character χ afforded by the $\mathcal{S}_n$-module M is called a *noncommutative character* corresponding to χ, or M.

* * *

By recourse to the noncommutative theory, the ordinary character theory of the symmetric group can be deduced by means of simple *noncommutative computations* in the coplactic algebra. Many classical results serve as examples in the third part of this book.

To start with, we shall consider the class functions $\zeta^p := c(\mathrm{Z}^p)$ of $\mathcal{S}_n$ indexed by partitions p of n. The Main Theorem together with the noncommutative orthogonality relations (1.3) implies that $\{\, \zeta^p \,|\, p \vdash n \,\}$ is an orthonormal basis of $\mathcal{C}\ell_K(\mathcal{S}_n)$. In Chapter 10, it will be shown that this is in fact the set of all irreducible characters of $\mathcal{S}_n$ and, in particular, that $\mathbb{Q}$ is a splitting field of $\mathcal{S}_n$. Accordingly, the sum Z^p of all standard Young tableaux of shape p in $K\mathcal{S}_n$ is a *noncommutative irreducible character* of $\mathcal{S}_n$, for each $p \vdash n$.

Let δ be a character of $\mathcal{S}_n$ with noncommutative counterpart Δ in $\mathcal{Q}$. If Δ is a sum of permutations, say, $\Delta = \sum_{\pi\in D} \pi$, then

$$(\delta, \zeta^p)_{\mathcal{C}} = (c(\Delta), c(\mathrm{Z}^p))_{\mathcal{C}} = (\Delta, \mathrm{Z}^p)_{\mathcal{P}} = |D^{-1} \cap \mathsf{SYT}^p|.$$

The scalar product on the left hand side yields the multiplicity of the irreducible character ζ^p in δ. Its noncommutative counterpart on the right hand side gives, by definition, a combinatorial description of this multiplicity, namely the number of standard Young tableaux $\pi \in \mathsf{SYT}^p$ such that $\pi^{-1} \in D$. This is a leading point for applications of noncommutative character theory.

For example, the sum $\Sigma\mathcal{S}_n = \sum_{\pi\in\mathcal{S}_n} \pi$ is an inverse image of the regular character $\chi_{K\mathcal{S}_n}$ of $\mathcal{S}_n$ under c and contained in $\mathcal{Q}$ — a noncommutative regular character. It follows without difficulty that

$$(\chi_{K\mathcal{S}_n}, \zeta^p)_{\mathcal{C}} = (\Sigma\mathcal{S}_n, \mathrm{Z}^p)_{\mathcal{P}} = \mathsf{syt}^p := |\mathsf{SYT}^p|.$$

In other words, the multiplicity of ζ^p in the regular $\mathcal{S}_n$-character is equal to the number of standard Young tableaux of shape p. But the same scalar product yields the degree $\deg \zeta^p$ of the irreducible character ζ^p. Hence, denoting by M_p an irreducible module affording the character ζ^p for all partitions p of n, we may conclude that $\dim M_p = \deg \zeta^p = \mathsf{syt}^p$ and

$$K\mathcal{S}_n \cong \bigoplus_{p \vdash n} \mathsf{syt}^p \, M_p \,. \tag{1.4}$$

This fundamental identity has as combinatorial refinement the decomposition

$$\Sigma\mathcal{S}_n = \sum_{p \vdash n} \sum_{\sigma \in \mathsf{SYT}^p} \Sigma A_\sigma$$

of $\mathcal{S}_n$ into coplactic classes. Indeed, $c(A_\sigma) = \zeta^p$ actually for any $\sigma \in \mathsf{SYT}^p$. When applying c, the latter equality thus turns into (1.4), expressed in terms of the corresponding characters.

In Chapters 11–13, few-line proofs are given in the same fashion of some classical results including the Littlewood–Richardson Theorem, the Branching Rule, Young's Rule and a combinatorial description of the Kostka numbers. Also without difficulty, the recursive formula for the irreducible character values known as the Murnaghan–Nakayama Rule can be stated and proved more generally as a formula for so-called skew characters, which reduces to the classical rule (due to Murnaghan and Nakayama) as a special case.

Further applications concern the descent characters $\delta^D = c(\Delta^D)$ indexed by subsets $D \subseteq \underline{n-1}$ in Chapter 14. To conclude, results of Kraskiewićz–Weyman [KW01] and Leclerc–Scharf–Thibon [LST96] are recovered in Chapter 15 on the cyclic characters of $\mathcal{S}_n$, which are induced from the cyclic subgroup generated by a long cycle.

PART I
The Inductive Method

Chapter 2

Coproducts

The classical representation theory of the symmetric group is distinguished by its combinatorial nature. Not surprisingly, major parts of the theory may be described elegantly by means of coproducts, or co- and bialgebras arising from these coproducts. We refer to Joni and Rota [JR82] who point out

> *that the notion of coalgebra, bialgebra, and Hopf algebra, recently introduced in mathematics, may give in a variety of cases a valuable formal framework for the study of combinatorial problems ... interesting both the combinatorist in search of a theoretical horizon, and the algebraist in search of examples which may point to new and general theorems.*

However, only those definitions needed here (and their illustration through examples) will be presented. Interested readers may find more details in [JR82; Mon93; Swe69].

Throughout this chapter, $(A, \star)$ is a K-Algebra. All tensor products are understood to be taken over the field K.

Observe that A is a K-vector space and $\star : A \times A \to A$ is a bilinear mapping. Linearising $\star$, we obtain a uniquely determined linear mapping

$$\mu : A \otimes A \longrightarrow A$$

such that $(x \otimes y)\mu = x \star y$ for all $x, y \in A$. Conversely, the multiplication $\star$ in A may clearly be recovered from μ. Dualising the linearised product μ, we define:

2.1 Definition. The pair (C, δ) is a *K-coalgebra* if C is a K-vector space and

$$\delta : C \to C \otimes C$$

is K-linear. In this case, the mapping δ is a *coproduct* on C.

2.2 Example. Let $K\mathbb{N}^*$ be a K-vector space with K-basis $\mathbb{N}^*$ and define a linear mapping $\mu : K\mathbb{N}^* \otimes K\mathbb{N}^* \to K\mathbb{N}^*$ by

$$(q \otimes r)\mu = q.r,$$

for all $q, r \in \mathbb{N}^*$. The corresponding multiplication turns $K\mathbb{N}^*$ into the ordinary semigroup algebra of the monoid $\mathbb{N}^*$ over K. Extending

$$\mathbb{N}^* \longrightarrow K\mathbb{N}^* \otimes K\mathbb{N}^*,\ q \longmapsto \sum_{\substack{u,v \in \mathbb{N}^* \\ u.v=q}} u \otimes v$$

linearly, there is a coproduct on $K\mathbb{N}^*$ which is denoted by Δ.

A *subalgebra* of A is a linear subspace T of A such that $T \star T \subseteq T$, that is

$$(T \otimes T)\mu \subseteq T.$$

Dualising again:

2.3 Definition. Let (C, δ) be a coalgebra. A linear subspace T of C is a *sub-coalgebra* of C if

$$T\delta \subseteq T \otimes T.$$

For example, the linear span of all partitions in $\mathbb{N}^*$ is a sub-coalgebra of the coalgebra $(K\mathbb{N}^*, \Delta)$ defined in 2.2.

Any common property of algebras may be translated into a suitable co-property of coalgebras. To give another example, let $\kappa_A : A \otimes A \to A \otimes A$ denote the linear mapping such that $(x \otimes y)\kappa_A = y \otimes x$, for all $x, y \in A$, and observe that A is a *commutative* algebra if

$$\kappa_A \mu = \mu.$$

This leads to:

2.4 Definition. A coalgebra (C, δ) is *cocommutative* if

$$\delta = \delta \kappa_C\ .$$

It is left to the reader to establish the notions of coassociativity, counit, and so on, since these concepts are not needed in what follows.

2.5 Example. As $(1.2)\Delta = \varnothing \otimes 1.2 + 1 \otimes 2 + 1.2 \otimes \varnothing$, the coalgebra $(K\mathbb{N}^*, \Delta)$ defined in example 2.2 is not cocommutative. A cocommutative coproduct δ on $K\mathbb{N}^*$ is obtained as follows: For any $q = q_1. \dots .q_k \in \mathbb{N}^*$ and any subset $J = \{j_1, \dots, j_l\}$ of $\{1, \dots, k\}$ such that $j_1 < j_2 < \dots < j_l$, put

$$q_J := q_{j_1}. \dots .q_{j_l}$$

and $\complement J := \underline{k} \backslash J$. Then the coproduct $\delta : K\mathbb{N}^* \to K\mathbb{N}^* \otimes K\mathbb{N}^*$ is defined by

$$q\delta := \sum_{J \subseteq \underline{k}} q_J \otimes q_{\complement J}$$

for all $q \in \mathbb{N}^*$, and linearity.

2.6 Definition and Remark. Let (C, δ) be a coalgebra and

$$\langle \cdot, \cdot \rangle : A \times C \to K$$

be a bilinear mapping. Let $\langle \cdot , \cdot \rangle_\otimes : (A \otimes A) \times (C \otimes C) \to K$ be the unique bilinear form such that

$$\langle a_1 \otimes a_2 , c_1 \otimes c_2 \rangle_\otimes = \langle a_1 , c_1 \rangle \langle a_2 , c_2 \rangle$$

for all $a_1, a_2 \in A$, $c_1, c_2 \in C$. Then the algebra A and the coalgebra C are called *dual with respect to* $\langle \cdot , \cdot \rangle$ if

$$\langle x \star y , z \rangle = \langle x \otimes y , z\delta \rangle_\otimes$$

for all $x, y \in A$, $z \in C$.

Note that the bilinear form $\langle \cdot , \cdot \rangle$ is not necessarily *regular*. Here regular means that $\langle a , c \rangle = 0$ for all $c \in C$ (respectively, $a \in A$) implies that $a = 0$ (respectively, $c = 0$). So, for instance, the algebra $(A, \star)$ is dual to *any* coalgebra (C, δ) with respect to the trivial form $A \times C \to K$ which maps every pair $(a, c) \in A \times C$ to zero.

2.7 Example. The algebra $(K\mathbb{N}^*, .)$ and the coalgebra $(K\mathbb{N}^*, \Delta)$ defined in 2.2 are dual with respect to that bilinear form $\langle \cdot , \cdot \rangle$ on $K\mathbb{N}^*$ which turns $\mathbb{N}^*$ into an orthonormal basis of $K\mathbb{N}^*$. Indeed, for any $q, r, s \in \mathbb{N}^*$,

$$\begin{aligned} \langle r \otimes s , q\Delta \rangle_\otimes &= \sum_{u.v=q} \langle r \otimes s , u \otimes v \rangle_\otimes \\ &= \sum_{u.v=q} \langle r , u \rangle \langle s , v \rangle \end{aligned}$$

$$= \begin{cases} 1 & \text{if } r.s = q \\ 0 & \text{otherwise} \end{cases}$$

$$= \langle r.s \, , q \rangle .$$

The notion of a bialgebra is crucial for the noncommutative approach to the character theory of the symmetric group.

2.8 Definition. Multiplying "component-wise", there is a product $\star_{\otimes}$ on $A \otimes A$ such that

$$(a_1 \otimes a_2) \star_{\otimes} (b_1 \otimes b_2) := (a_1 \star b_1) \otimes (a_2 \star b_2)$$

for all $a_1, a_2, b_1, b_2 \in A$. The linearisation $(A \otimes A) \otimes (A \otimes A) \to A \otimes A$ of this product is denoted by $\mu_{\otimes}$.

The triple $(A, \star, \delta)$ is a K-*bialgebra* if (A, δ) is a K-coalgebra such that δ is a homomorphism with respect to $\star$ in the following sense:

$$(x \star y)\delta = (x\delta) \star_{\otimes} (y\delta)$$

for all $x, y \in A$; that is, $\mu\delta = (\delta \otimes \delta)\mu_{\otimes}$.

As usual, a linear subspace B of a bialgebra A is a *sub-bialgebra* if B is a subalgebra and a sub-coalgebra of A.

2.9 Example. $(K\mathbb{N}^*, \, . \, , \delta)$ is a bialgebra, where δ is the cocommutative coproduct on $K\mathbb{N}^*$ defined in 2.5. For, if $q = q_1. \dots .q_k, r = r_1. \dots .r_l \in \mathbb{N}^*$ and $n := k + l$, then

$$\begin{aligned}
(q.r)\delta &= \sum_{J \subseteq \underline{n}} (q.r)_J \otimes (q.r)_{\complement J} \\
&= \sum_{J_1 \subseteq \underline{k}} \ \sum_{J_2 \subseteq \underline{n} \setminus \underline{k}} (q.r)_{J_1 \cup J_2} \otimes (q.r)_{\complement(J_1 \cup J_2)} \\
&= \sum_{J_1 \subseteq \underline{k}} \ \sum_{J_2 \subseteq \underline{l}} q_{J_1} . r_{J_2} \otimes q_{\complement J_1} . r_{\complement J_2} \\
&= \Big(\sum_{J_1 \subseteq \underline{k}} q_{J_1} \otimes q_{\complement J_1} \Big) \cdot_{\otimes} \Big(\sum_{J_2 \subseteq \underline{l}} r_{J_2} \otimes r_{\complement J_2} \Big) \\
&= q\delta \cdot_{\otimes} r\delta \, .
\end{aligned}$$

Note that δ is the unique coproduct on $K\mathbb{N}^*$ such that $(K\mathbb{N}^*, ., \delta)$ is a bialgebra and $n\delta = n \otimes \varnothing + \varnothing \otimes n$ for all $n \in \mathbb{N}$, since $\mathbb{N}$ generates $K\mathbb{N}^*$ as an algebra.

In 9.7, we shall see that the bialgebra $(K\mathbb{N}^*, ., \delta)$ is isomorphic to the bialgebra $(\mathcal{D}, \star, \downarrow)$ defined in the introduction.

2.10 Definition. Let $(A, \star, \delta)$, $(B, \bullet, \Delta)$ be bialgebras. Let

$$\langle \cdot, \cdot \rangle : A \times B \to K$$

be bilinear and define $\tau : B \times A \to A \times B, (b, a) \mapsto (a, b)$. Then A and B are said to be *dual* with respect to $\langle \cdot, \cdot \rangle$ if $(A, \star)$ is dual to (B, Δ) with respect to $\langle \cdot, \cdot \rangle$ and $(B, \bullet)$ is dual to (A, δ) with respect to $\tau\langle \cdot, \cdot \rangle$. When $(A, \star, \delta) = (B, \bullet, \Delta)$, then A is called *self-dual* with respect to $\langle \cdot, \cdot \rangle$. In particular, if $\langle \cdot, \cdot \rangle$ is a symmetric bilinear form on A, then A is self-dual if and only if

$$\langle a_1 \star a_2 , a_3 \rangle = \langle a_1 \otimes a_2 , a_3\delta \rangle_{\otimes}$$

for all $a_1, a_2, a_3 \in A$.

2.11 Definition. Let $n \in \mathbb{N}$ and $q = q_1. \dots .q_l \models n$, then $\ell(q) := l$ is the *length* of q. We set

$$q? := q_1 \cdots q_l \, m_1! \cdots m_n! \in \mathbb{N},$$

where m_i denotes the multiplicity of the letter i in q, for all $i \in \underline{n}$.

Furthermore, $r \in \mathbb{N}^*$ is a *rearrangement* of q if $\ell(q) = \ell(r)$ and there exists a permutation $\pi \in \mathcal{S}_l$ such that $r = q_{1\pi}. \dots .q_{l\pi}$. In this case, write $r \approx q$. In other words, $r \approx q$ if and only if r may be obtained by rearranging the letters of q.

2.12 Example. The bialgebra $(K\mathbb{N}^*, ., \delta)$ considered in 2.9 is self-dual with respect to the bilinear mapping $(\cdot, \cdot)$, defined by

$$(q, r) := \begin{cases} q? & \text{if } q \approx r \\ 0 & \text{otherwise} \end{cases}$$

for all $q, r \in \mathbb{N}^*$. Indeed, there is the perfect analogue of the binomial coefficient

$$\frac{(r.s)?}{r?\, s?} = \#\{\, J \subseteq \underline{\ell(r.s)} \mid (r.s)_J \approx r \,\}$$

for all $r, s \in \mathbb{N}^*$. Hence

$$(r \otimes s, q\delta)_{\otimes} = \sum_{J \subseteq \underline{k}} (r, q_J)(s, q_{\complement J})$$

$$= \begin{cases} \#\{\, J \mid q_J \approx r \,\} r?\, s? & \text{if } q \approx r.s \\ 0 & \text{if } q \not\approx r.s \end{cases}$$

$$= \begin{cases} (r.s)? & \text{if } q \approx r.s \\ 0 & \text{if } q \not\approx r.s \end{cases}$$

$$= (r.s, q)$$

for all $q, r, s \in \mathbb{N}^*$.

If $(B, \bullet)$ is a K-algebra, then a *homomorphism of algebras* from A to B is a linear map $\varphi : A \to B$ such that $(a \star b)\varphi = (a\varphi) \bullet (b\varphi)$ for all $a, b \in A$, that is,

$$\mu_\star \varphi = (\varphi \otimes \varphi)\mu_\bullet \,.$$

2.13 Definition. Let (C, δ) and (B, Δ) be coalgebras, then a *homomorphism of coalgebras* from C to B is a linear map $\varphi : C \to B$ such that

$$\varphi\Delta = \delta(\varphi \otimes \varphi),$$

that is: $(c\varphi)\Delta = (c\delta)(\varphi \otimes \varphi)$ for all $c \in C$.

Of course, a *homomorphism of bialgebras* is a simultaneous homomorphism of algebras and coalgebras.

An illustration of the formalism introduced so far follows which will be of use in Chapter 9. Recall that a linear map φ from a K-vector space A with bilinear form $\langle \cdot\,, \cdot \rangle_A$ to a K-vector space B with bilinear form $\langle \cdot\,, \cdot \rangle_B$ is *isometric* if

$$\langle a_1\,, a_2 \rangle_A = \langle a_1\varphi\,, a_2\varphi \rangle_B$$

for all $a_1, a_2 \in A$.

2.14 Proposition. *Let $(A, \star, \delta)$ and $(B, \bullet, \Delta)$ be self-dual bialgebras and suppose the underlying bilinear form on B is regular, then each isometric epimorphism of algebras from A onto B is an isometric epimorphism of bialgebras.*

Proof. Denote the bilinear forms on A and B by $\langle \cdot\,, \cdot \rangle_A$ and $\langle \cdot\,, \cdot \rangle_B$, respectively. Let φ be an isometric homomorphism of algebras from A to B and assume that φ is onto.

Let $a \in A$ and $b_1, b_2 \in B$ and choose $a_1, a_2 \in A$ such that $a_1\varphi = b_1$ and $a_2\varphi = b_2$ to obtain

$$\begin{aligned}
\Big\langle (a\delta)(\varphi \otimes \varphi)\,,\, b_1 \otimes b_2 \Big\rangle_{B,\otimes} &= \Big\langle (a\delta)(\varphi \otimes \varphi)\,,\, a_1\varphi \otimes a_2\varphi \Big\rangle_{B,\otimes} \\
&= \Big\langle (a\delta)(\varphi \otimes \varphi)\,,\, (a_1 \otimes a_2)(\varphi \otimes \varphi) \Big\rangle_{B,\otimes} \\
&= \langle a\delta\,,\, a_1 \otimes a_2 \rangle_{A,\otimes} \\
&= \langle a\,,\, a_1 \star a_2 \rangle_A \\
&= \Big\langle a\varphi\,,\, (a_1 \star a_2)\varphi \Big\rangle_B \\
&= \Big\langle a\varphi\,,\, a_1\varphi \bullet a_2\varphi \Big\rangle_B \\
&= \Big\langle (a\varphi)\Delta\,,\, b_1 \otimes b_2 \Big\rangle_{B,\otimes}\,.
\end{aligned}$$

Regularity of $\langle\,\cdot\,,\,\cdot\,\rangle_B$ thus implies the homomorphism rule for the coproducts. □

Chapter 3

The Bialgebra $\mathcal{C}$ of Class Functions

Basic elements of character theory can be gathered together with the specific properties of symmetric groups arising from their inductive structure and put to use in the bialgebra of class functions introduced by Geissinger [Gei77].

As indicated in the introduction, the notions of induction and restriction of class functions naturally yield a product $\bullet$, and a coproduct $\downarrow$, respectively, on the direct sum

$$\mathcal{C} := \bigoplus_{n \in \mathbb{N}_0} \mathcal{C}\ell_K(\mathcal{S}_n)$$

turning it into a commutative bialgebra. Induction and restriction are intimately linked by Frobenius' reciprocity law which, in the language of bialgebras, occurs as a self-duality of $\mathcal{C}$, once the scalar products $(\cdot,\cdot)_{\mathcal{S}_n}$ on $\mathcal{C}\ell_K(\mathcal{S}_n)$, $n \in \mathbb{N}_0$, are properly extended to $\mathcal{C}$.

Due to Frobenius, the algebra $(\mathcal{C}, \bullet)$ is isomorphic to the *algebra of symmetric functions* Λ over K, which is therefore equally suitable to encode the commutative part of the theory (see 3.8). The standard reference for the theory of symmetric functions is Macdonald's monograph [Mac95].

Let us start with the construction of the bialgebra structure on $\mathcal{C}$. A concrete realisation of the direct product $\mathcal{S}_k \times \mathcal{S}_l$ ($k, l \in \mathbb{N}_0$) as a subgroup of $\mathcal{S}_{k+l}$ is obtained as follows.

3.1 Notation and Remarks. Let $k, l \in \mathbb{N}_0$ and put $n := k + l$. For any $\pi \in \mathcal{S}_k$ and $\sigma \in \mathcal{S}_l$, let $\pi\#\sigma$ be the permutation in $\mathcal{S}_n$ acting on $\underline{k}$ according to π, and on $\underline{n} \setminus \underline{k}$ according to σ. More formally, define $\pi\#\sigma$ by

$$i(\pi\#\sigma) := \begin{cases} i\pi & \text{if } i \leq k, \\ (i-k)\sigma + k & \text{if } i > k, \end{cases}$$

for all $i \in \underline{n_l}$. Then $\mathcal{S}_{k.l} = \mathcal{S}_k \# \mathcal{S}_l$ is the Young subgroup of $\mathcal{S}_n$ of type $k.l$ and the mapping

$$\mathcal{S}_k \times \mathcal{S}_l \to \mathcal{S}_{k.l}, \ (\pi, \sigma) \mapsto \pi \# \sigma$$

is an isomorphism of groups.

3.2 Definition and Remark. Let $k, l \in \mathbb{N}_0$. For all $\alpha \in \mathcal{C}\ell_K(\mathcal{S}_k)$ and $\beta \in \mathcal{C}\ell_K(\mathcal{S}_l)$, define $\alpha \# \beta : \mathcal{S}_{k.l} \to K$ by

$$(\alpha \# \beta)(\pi \# \sigma) := \alpha(\pi)\, \beta(\sigma),$$

for all $\pi \in \mathcal{S}_k$ and $\sigma \in \mathcal{S}_l$. Then $\alpha \# \beta \in \mathcal{C}\ell_K(\mathcal{S}_{k.l})$. Now set

$$\alpha \bullet \beta := (\alpha \# \beta)^{\mathcal{S}_{k+l}},$$

the class function of $\mathcal{S}_{k+l}$ induced by $\alpha \# \beta$ (see A.2.8 in Appendix A). Then $\bullet$ is linear in both components. By bilinear extension, we obtain the *outer product* $\bullet$ on $\mathcal{C} = \bigoplus_{n \geq 0} \mathcal{C}\ell_K(\mathcal{S}_n)$.

For example, if $q \models n$ and $r \models m$, then there is the multiplication rule

$$\begin{aligned} \xi^r \bullet \xi^q &= \left((1_{\mathcal{S}_r})^{\mathcal{S}_m} \# (1_{\mathcal{S}_q})^{\mathcal{S}_n}\right)^{\mathcal{S}_{n+m}} \\ &= \left((1_{\mathcal{S}_r} \# 1_{\mathcal{S}_q})^{\mathcal{S}_{m.n}}\right)^{\mathcal{S}_{n+m}} \\ &= (1_{\mathcal{S}_{r.q}})^{\mathcal{S}_{n+m}} \\ &= \xi^{r.q} \end{aligned}$$

for the corresponding Young characters, by transitivity of induction (see A.2.8).

3.3 Proposition. *Let $k, l \in \mathbb{N}_0$, then for any character χ of $\mathcal{S}_k$ and any character ψ of $\mathcal{S}_l$, the outer product $\chi \bullet \psi$ is a character of $\mathcal{S}_{k+l}$.*

Proof. Let M be an $\mathcal{S}_k$-module affording χ, and let N be an $\mathcal{S}_l$-module affording ψ, then $M \otimes N$ is an $\mathcal{S}_k \times \mathcal{S}_l$-module, hence also an $\mathcal{S}_{k.l}$-module, by 3.1. The character of $\mathcal{S}_{k.l}$ afforded by $M \otimes N$ is readily seen to be $\chi \# \psi$. Thus $\chi \bullet \psi$ is induced by the character $\chi \# \psi$ of $\mathcal{S}_{k.l}$, hence itself a character (see A.2.10). □

3.4 Notation and Remarks. Let $n \in \mathbb{N}_0$ and $q \models n$. Denote by C_q the conjugacy class in $\mathcal{S}_n$ consisting of all permutations π in $\mathcal{S}_n$ with cycle type obtained by rearranging q. Then $C_q^{-1} = \{\pi^{-1} \mid \pi \in C_q\} = C_q$ and $C_r = C_q$

for any rearrangement r of q.* In particular, the number of conjugacy classes in $\mathcal{S}_n$ equals the number of partitions of n. If $\alpha \in \mathcal{C}\ell_K(\mathcal{S}_n)$, it is convenient to denote the unique value of α on any element $\pi \in C_q$ by $\alpha(C_q)$, for all $q \models n$. In fact, we shall sometimes consider $\alpha = \sum_{n\in\mathbb{N}_0} \alpha_n \in \mathcal{C}$ with $\alpha_n \in \mathcal{C}\ell_K(\mathcal{S}_n)$ for all $n \in \mathbb{N}_0$ and write $\alpha(C_q) = \alpha_n(C_q)$ for all $q \models n$, by abuse of notation.

We leave it to the reader to verify that the order of the centraliser $C_{\mathcal{S}_n}(\pi)$ in $\mathcal{S}_n$ of an element $\pi \in C_q$ is equal to the number $q?$ defined in 2.11, that is,

$$q? = n!/|C_q|.$$

3.5 Notation and Remarks. Let char_q denote the characteristic function of C_q in $\mathcal{C}\ell_K(\mathcal{S}_n)$, mapping $\pi \in \mathcal{S}_n$ to one or zero according as $\pi \in C_q$ or not. Then $\mathsf{char}_q = \mathsf{char}_r$ for those $q, r \models n$ such that $q \approx r$ and $\{\, \mathsf{char}_p \,|\, p \vdash n \,\}$ is a K-basis of $\mathcal{C}\ell_K(\mathcal{S}_n)$. Using suitable scalar multiples of the characteristic functions, namely

$$\mathsf{ch}_q := q?\mathsf{char}_q$$

for all $q \in \mathbb{N}^*$, there is a linear basis $\{\, \mathsf{ch}_q \,|\, q \in \mathbb{N}^* \,\}$ of $\mathcal{C}$ with the following useful property:

3.6 Proposition. *Let $n \in \mathbb{N}_0$, then*

$$(\alpha, \mathsf{ch}_q)_{\mathcal{S}_n} = \alpha(C_q),$$

for all $q \models n$ and $\alpha \in \mathcal{C}\ell_K(\mathcal{S}_n)$.

Proof. By definition, $(\alpha, \mathsf{ch}_q)_{\mathcal{S}_n} = 1/n! \sum_{\pi\in C_q} q?\alpha(\pi) = \alpha(C_q)$. □

The structure of the algebra $(\mathcal{C}, \bullet)$ has a simple description in terms of this basis.

3.7 Theorem. *$(\mathcal{C}, \bullet)$ is an associative and commutative algebra. Furthermore, for all $q, r \in \mathbb{N}^*$,*

$$\mathsf{ch}_q \bullet \mathsf{ch}_r = \mathsf{ch}_{q.r}\,.$$

In other words, $\mathcal{C}$ is a ring of polynomials in the set of (commuting) variables ch_n, $n \in \mathbb{N}$.

---

*This redundance will be advantageous at a later stage (see, for instance, 3.13).

Proof. Let $k, l \in \mathbb{N}_0$, $n = k + l$, $q \models k$ and $r \models l$, then, by definition of the outer product,

$$(\mathsf{ch}_q \bullet \mathsf{ch}_r)(\pi) = \frac{1}{k!l!} \sum_{\substack{\alpha \in \mathcal{S}_n \\ \alpha^{-1}\pi\alpha \in \mathcal{S}_{k.l}}} (\mathsf{ch}_q \# \mathsf{ch}_r)(\alpha^{-1}\pi\alpha)$$

$$= \frac{q?r?}{k!l!} \left| \left\{ \alpha \in \mathcal{S}_n \,\middle|\, \alpha^{-1}\pi\alpha \in C_q \# C_r \right\} \right|$$

for each $\pi \in \mathcal{S}_n$. This is zero unless $\pi \in C_{q.r}$, since $C_q \# C_r \subseteq C_{q.r}$. If $\pi \in C_{q.r}$, then $(\mathsf{ch}_q \bullet \mathsf{ch}_r)(\pi) = \frac{1}{k!l!} q?r? \, |C_q \# C_r| |C_{\mathcal{S}_n}(\pi)| = (q.r)?$. This proves the multiplication rule. All the remaining claims follow, as the elements ch_p indexed by partitions $p \in \mathbb{N}^*$ constitute a linear basis of $\mathcal{C}$. □

3.8 Remark. The preceding result allows us to define Frobenius' characteristic mapping from $\mathcal{C}$ to the algebra of symmetric functions Λ over K, as follows.

The algebra Λ is a subalgebra of the algebra of formal power series over K in infinitely many variables $x_1, x_2, \ldots$. More precisely, the symmetric power sums p_n, defined by

$$p_n = \sum_{k \geq 1} x_k^n = x_1^n + x_2^n + \cdots$$

for all $n \in \mathbb{N}$, are algebraically independent over K and form a set of algebra generators of Λ (see [Mac95, (2.12), p. 24]). Thus, by the above theorem, the *Frobenius characteristic* $F : \mathcal{C} \to \Lambda$, defined by $\mathsf{ch}_n \mapsto p_n$ for all $n \in \mathbb{N}$, is an isomorphism of algebras.

The image of the trivial $\mathcal{S}_n$-character under this isomorphism is the *complete symmetric function* h_n, which is the sum of all monomials in $x_1, x_2, \ldots$ of degree n, while the sign character of $\mathcal{S}_n$ (see 10.6(ii)) is mapped to the *elementary symmetric function* $e_n = \sum_{i_1 < i_2 < \cdots < i_n} x_{i_1} x_{i_2} \cdots x_{i_n}$.

The Frobenius characteristic is the link between the representation theory of the symmetric group and of the general linear group as discovered by Schur in his dissertation [Sch01] and a famous subsequent paper [Sch27]. As a part of this, the symmetric function corresponding to an irreducible $\mathcal{S}_n$-character χ was determined and is now referred to as the *Schur function* corresponding to χ (see [Mac95, I, 7]).

The coproduct on $\mathcal{C}$ arises from the restriction of class functions.

3.9 Definition and Remarks. Canonically,

$$\mathcal{C}\otimes\mathcal{C} = \Big(\bigoplus_{n\in\mathbb{N}_0} \mathcal{C}\ell_K(\mathcal{S}_n)\Big)\otimes\Big(\bigoplus_{n\in\mathbb{N}_0} \mathcal{C}\ell_K(\mathcal{S}_n)\Big) \cong \bigoplus_{k,l\in\mathbb{N}_0}\Big(\mathcal{C}\ell_K(\mathcal{S}_k)\otimes\mathcal{C}\ell_K(\mathcal{S}_l)\Big).$$

If $k,l \in \mathbb{N}_0$, then the set $\{\,\mathsf{char}_q \otimes \mathsf{char}_r \mid q \vdash k,\, r \vdash l\,\}$ is a linear basis of $\mathcal{C}\ell_K(\mathcal{S}_k)\otimes\mathcal{C}\ell_K(\mathcal{S}_l)$, while the elements $\mathsf{char}_q \# \mathsf{char}_r$ $(q \vdash k,\, r \vdash l)$ constitute a linear basis of $\mathcal{C}\ell_K(\mathcal{S}_{k.l})$, by 3.5. Denote by

$$i_{k,l} : \mathcal{C}\ell_K(\mathcal{S}_k)\otimes\mathcal{C}\ell_K(\mathcal{S}_l) \to \mathcal{C}\ell_K(\mathcal{S}_{k.l})$$

the linear isomorphism which takes $\mathsf{char}_q \otimes \mathsf{char}_r \mapsto \mathsf{char}_q \# \mathsf{char}_r$ for all $q \vdash k$, $r \vdash l$ and define linear mappings

$$\downarrow_n : \mathcal{C}\ell_K(\mathcal{S}_n) \to \bigoplus_{k+l=n} \mathcal{C}\ell_K(\mathcal{S}_k)\otimes\mathcal{C}\ell_K(\mathcal{S}_l), \quad \alpha\downarrow_n := \sum_{k+l=n} \alpha|_{\mathcal{S}_{k.l}} i_{k,l}^{-1}$$

for all $n \in \mathbb{N}_0$. The coproduct $\downarrow : \mathcal{C} \to \mathcal{C}\otimes\mathcal{C}$ is then defined as the unique common linear extension of all $\downarrow_n$, $n \in \mathbb{N}_0$.

For example, for the trivial character ξ^n of $\mathcal{S}_n$, there is the coproduct rule

$$\xi^n\downarrow = \sum_{k=0}^{n} \xi^k \otimes \xi^{n-k},$$

since the restriction of ξ^n to the subgroup $\mathcal{S}_{k.(n-k)}$ is the trivial character $\xi^k \# \xi^{n-k}$ of $\mathcal{S}_{k.(n-k)}$ for all $k \in \underline{n} \cup \{0\}$.

3.10 Lemma. *$(\mathcal{C},\downarrow)$ is a cocommutative coalgebra. Furthermore,*

$$\mathsf{ch}_q\downarrow = \sum_{J\subseteq \underline{k}} \mathsf{ch}_{q_J} \otimes \mathsf{ch}_{q_{CJ}},$$

for all $q = q_1.\dots.q_k \in \mathbb{N}^*$.

Proof. Let $a,b \in \mathbb{N}_0$, $n := a+b$ and $q = q_1.\dots.q_k \models n$, then the intersection of C_q with $\mathcal{S}_{a.b}$ is the union of all conjugacy classes $C_r \# C_s$ of $\mathcal{S}_{a.b}$ contained in C_q. The binomial formula for $(r.s)?/(r?s?)$ mentioned in 2.12 gives

$$\begin{aligned}\mathsf{ch}_q|_{\mathcal{S}_{a.b}} &= q?\mathsf{char}_q|_{\mathcal{S}_{a.b}} \\ &= q? \underbrace{\sum_{r\vdash a}\sum_{s\vdash b}}_{C_r\#C_s\subseteq C_q} \mathsf{char}_r \# \mathsf{char}_s\end{aligned}$$

$$= q? \underbrace{\sum_{r \vdash a} \sum_{s \vdash b}}_{r.s \approx q} \mathsf{char}_r \# \mathsf{char}_s$$

$$= \underbrace{\sum_{r \vdash a} \sum_{s \vdash b}}_{r.s \approx q} \frac{(r.s)?}{r?s?} \mathsf{ch}_r \# \mathsf{ch}_s$$

$$= \underbrace{\sum_{r \vdash a} \sum_{s \vdash b} \sum_{J \subseteq \underline{k}}}_{q_J \approx r,\ q_{CJ} \approx s} \mathsf{ch}_{q_J} \# \mathsf{ch}_{q_{CJ}}$$

$$= \sum_{\substack{J \subseteq \underline{k} \\ q_J \models a}} \mathsf{ch}_{q_J} \# \mathsf{ch}_{q_{CJ}} \,.$$

Taking inverse images under $i_{a,b}$ and summing over all possible values of a and b such that $a + b = n$ yields the asserted formula for the coproducts. Cocommutativity is now readily seen. □

In order to complete the picture, consider the orthogonal extension of the standard scalar products on $\mathcal{C}\ell_K(\mathcal{S}_n)$, $n \in \mathbb{N}_0$:

3.11 Notation and Remarks. Define

$$(\alpha, \beta)_{\mathcal{C}} := \begin{cases} (\alpha, \beta)_{\mathcal{S}_k} & \text{if } k = l, \\ 0 & \text{if } k \neq l, \end{cases}$$

for all $k, l \in \mathbb{N}_0$ and all $\alpha \in \mathcal{C}\ell_K(\mathcal{S}_k)$, $\beta \in \mathcal{C}\ell_K(\mathcal{S}_l)$. This gives rise to a regular and symmetric bilinear form

$$(\,\cdot\,,\,\cdot\,)_{\mathcal{C}} : \mathcal{C} \times \mathcal{C} \to K$$

on $\mathcal{C} = \bigoplus_{n \in \mathbb{N}_0} \mathcal{C}\ell_K(\mathcal{S}_n)$, by bilinearity. Furthermore, denote the unique bilinear form on $\mathcal{C} \otimes \mathcal{C}$ inherited from $(\,\cdot\,,\,\cdot\,)_{\mathcal{C}}$ (as described in 2.6) by $(\,\cdot\,,\,\cdot\,)_{\mathcal{C} \otimes \mathcal{C}}$, then

$$(\alpha_1 \otimes \alpha_2, \beta_1 \otimes \beta_2)_{\mathcal{C} \otimes \mathcal{C}} = (\alpha_1, \beta_1)_{\mathcal{C}} (\alpha_2, \beta_2)_{\mathcal{C}}$$

for all $\alpha_1, \alpha_2, \beta_1, \beta_2 \in \mathcal{C}$.

Frobenius' reciprocity law (see A.2.9) for all symmetric groups $G = \mathcal{S}_n$ and all subgroups $U = \mathcal{S}_{k.l}$, $k + l = n$, yields:

3.12 Reciprocity Law. $(\alpha \bullet \beta, \gamma)_{\mathcal{C}} = (\alpha \otimes \beta, \gamma \downarrow)_{\mathcal{C} \otimes \mathcal{C}}$, *for all* $\alpha, \beta, \gamma \in \mathcal{C}$.

Proof. Let $k, l, n \in \mathbb{N}_0$ such that $n = k + l$ and let α, β and γ be class functions of $\mathcal{S}_k$, $\mathcal{S}_l$ and $\mathcal{S}_n$, respectively, then

$$\begin{aligned}
(\alpha \otimes \beta, \gamma\downarrow)_{\mathcal{C}\otimes\mathcal{C}} &= \Big(\alpha \otimes \beta,\ \sum_{a+b=n} \gamma|_{\mathcal{S}_{a.b}} i_{a,b}^{-1}\Big)_{\mathcal{C}\otimes\mathcal{C}} \\
&= \Big(\alpha \otimes \beta,\ \gamma|_{\mathcal{S}_{k.l}} i_{k,l}^{-1}\Big)_{\mathcal{C}\otimes\mathcal{C}} \\
&= \Big(\alpha \# \beta,\ \gamma|_{\mathcal{S}_{k.l}}\Big)_{\mathcal{S}_{k.l}} \\
&= \Big((\alpha \# \beta)^{\mathcal{S}_n},\ \gamma\Big)_{\mathcal{S}_n} \\
&= (\alpha \bullet \beta, \gamma)_{\mathcal{C}} .
\end{aligned}$$

□

Note that 3.6 implies

$$(\mathsf{ch}_q, \mathsf{ch}_r)_{\mathcal{C}} = \begin{cases} q? & \text{if } q \approx r \\ 0 & \text{otherwise} \end{cases}$$

for all $q, r \in \mathbb{N}^*$. Therefore, considering basis elements $\alpha = \mathsf{ch}_r$, $\beta = \mathsf{ch}_s$, $\gamma = \mathsf{ch}_q$ of $\mathcal{C}$, it follows that an alternative proof of 3.12 was given in 2.12 already.

We summarise this chapter with the following

3.13 Theorem. *$(\mathcal{C}, \bullet, \downarrow)$ is a commutative and cocommutative graded bialgebra and self-dual with respect to $(\cdot,\cdot)_{\mathcal{C}}$. For all $q = q_1.\dots.q_k, r \in \mathbb{N}^*$,*

$$\mathsf{ch}_q \bullet \mathsf{ch}_r = \mathsf{ch}_{q.r}$$

and

$$\mathsf{ch}_q\downarrow = \sum_{J \subseteq \underline{k}} \mathsf{ch}_{q_J} \otimes \mathsf{ch}_{q_{CJ}} .$$

Herein, the fact that $(\mathcal{C}, \bullet, \downarrow)$ is a bialgebra is an immediate consequence of 2.9, since the linear map $(K\mathbb{N}^*, ., \delta) \to (\mathcal{C}, \bullet, \downarrow)$, defined by $q \mapsto \mathsf{ch}_q$ for all $q \in \mathbb{N}^*$, is a homomorphism of algebras and of coalgebras.

PART II

Noncommutative Character Theory of the Symmetric Group

Chapter 4

Shapes

The notion of a *shape* is inspired by Stanley's theory of *P*-partitions [Sta72]. However, the study of partial orders on a finite set here should be viewed as a helpful tool for a transparent approach to the noncommutative bialgebras introduced in the chapters that follow, rather than an introduction to this field of combinatorics. The main goal of this chapter will be to make available an inductive technique (appearing in Lemma 4.11) which was introduced by Gessel in [Ges84].

In what follows, $\leq$ denotes the usual order on the set $\mathbb{Z}$ of integers.

4.1 Definition. Let S be a finite set, $\rightarrow_S$ be a total order and $\leq_S$ be a partial order on S. Then the triple

$$(S, \rightarrow_S, \leq_S)$$

is a *shape*, often simply written as S instead of $(S, \rightarrow_S, \leq_S)$. If $n = |S|$, then there is a unique order respecting bijection

$$\iota_S : (\underline{n}, \leq) \longrightarrow (S, \rightarrow_S),$$

the *natural labeling* of S. Two shapes S and T are *isomorphic* if there exists a bijection $\varphi : S \rightarrow T$ which is an isomorphism of ordered sets from $(S, \leq_S)$ to $(T, \leq_T)$ and from $(S, \rightarrow_S)$ to $(T, \rightarrow_T)$. We write $S \simeq T$ in this case.

4.2 Remark. Let S be a shape and $n := |S|$, then the natural labeling ι_S of S yields an isomorphism of shapes

$$(\underline{n}, \leq, \preceq_S) \longrightarrow (S, \rightarrow_S, \leq_S),$$

where the partial order $\preceq_S$ on $\underline{n}$ is defined by $i \preceq_S j$ if and only if $i\iota_S \leq_S j\iota_S$, for all $i, j \in \underline{n}$. However, in some situations, when studying combinatorial properties of shapes, the focus on partial orders of $\underline{n}$ turns out to be inconvenient and restrictive.

4.3 Special case. Let S be a shape and assume that $\leq_S$ is a *total* order on S. Let $n := |S|$, then there is a unique order respecting bijection φ from $(S, \leq_S)$ onto $(\underline{n}, \leq)$, and $\pi := \iota_S\varphi \in \mathcal{S}_n$. The shape S is isomorphic to the π-*shape*

$$S(\pi) := (\underline{n}, \rightarrow_\pi, \leq_\pi),$$

with total and partial order defined by

$$i \rightarrow_\pi j :\Longleftrightarrow i \leq j \quad \text{and} \quad i \leq_\pi j :\Longleftrightarrow i\pi \leq j\pi$$

for all $i, j \in \underline{n}$. The natural labeling of the π-shape is $\iota_{S(\pi)} = \mathsf{id}_n$.

Proof. This follows directly from the above remark. Indeed, for all $i, j \in \underline{n}$, $i \preceq_S j$ means $i\iota_S \leq_S j\iota_S$, by definition, hence $i\pi \leq j\pi$ after application of φ. This implies $S(\pi) = (\underline{n}, \leq, \preceq_S) \simeq S$. □

An important class of examples is provided by shapes which are contained in $\mathbb{Z} \times \mathbb{Z}$ and equipped with orders defined as follows.

4.4 Definition and Remark. For all $(i,j), (k,l) \in \mathbb{Z} \times \mathbb{Z}$, we define

$$(i,j) \leq_{\mathbb{Z}\times\mathbb{Z}} (k,l) :\Longleftrightarrow i \leq k \text{ and } j \leq l$$

and

$$(i,j) \rightarrow (k,l) :\Longleftrightarrow i > k \text{ or } (i = k \text{ and } j \leq l).$$

Then $\leq_{\mathbb{Z}\times\mathbb{Z}}$ is a partial and $\rightarrow$ is a total order on $\mathbb{Z} \times \mathbb{Z}$.

Any finite subset $S \subseteq \mathbb{Z} \times \mathbb{Z}$ can thus be viewed as a shape, with total and partial order inherited from $\rightarrow$ and $\leq_{\mathbb{Z}\times\mathbb{Z}}$, respectively. A more detailed analysis of shapes essentially of this kind follows in Chapter 6. For the time being, they may be viewed as a nice tool to illustrate the abstract definitions. Consider the following picture: Any element $x = (i,j) \in \mathbb{Z} \times \mathbb{Z}$ is called a *cell* and will be illustrated by a square box. The first component i of x gives the *row* of $\mathbb{Z} \times \mathbb{Z}$ containing x, with the understanding that the i-th row lies below the $(i+1)$-th row. Similarly, the second component j of x gives the *column* of $\mathbb{Z} \times \mathbb{Z}$ containing x. Here the j-th column lies to the left of the $(j+1)$-th column. For example, the picture corresponding to the five cells of the shape $S = \{(1,1), (1,2), (1,3), (2,1), (2,2)\}$ is

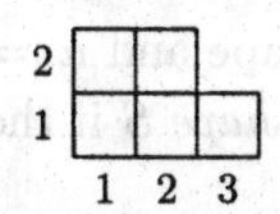

For $x \in \mathbb{Z} \times \mathbb{Z}$, insert a diagonal of negative slope ($\diagdown$) into the cells $y \in F \setminus \{x\}$ such that $x \leq_{\mathbb{Z}\times\mathbb{Z}} y$, and a diagonal of positive slope ($\diagup$) into the cells $y \in F \setminus \{x\}$ such that $x \to y$ to get the following helpful illustration of 4.4:

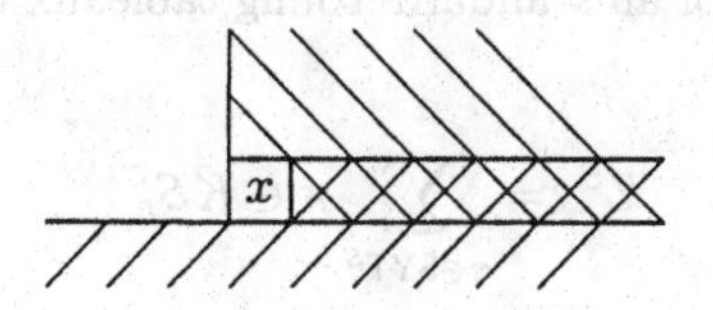

4.5 Proposition. *For all $x, y, z \in \mathbb{Z} \times \mathbb{Z}$,*

$$x \leq_{\mathbb{Z}\times\mathbb{Z}} y \iff x + z \leq_{\mathbb{Z}\times\mathbb{Z}} y + z$$

and

$$x \to y \iff (x + z) \to (y + z).$$

In other words, for each $z \in \mathbb{Z} \times \mathbb{Z}$, the mapping $\mathbb{Z} \times \mathbb{Z} \to \mathbb{Z} \times \mathbb{Z}$, $x \mapsto x + z$ is an isomorphism of ordered sets for both orders $\leq_{\mathbb{Z}\times\mathbb{Z}}$ and $\to$.

This is immediate from the definition.

4.6 Example. Let $S \subseteq \mathbb{Z} \times \mathbb{Z}$ be finite, then S is a shape with total and partial order inherited from $\to$ and $\leq_{\mathbb{Z}\times\mathbb{Z}}$, respectively, as described above. By 4.5, it is not necessary to fix the offset of S in $\mathbb{Z} \times \mathbb{Z}$, up to isomorphism of shapes. For example, let us say that the shape S is illustrated by

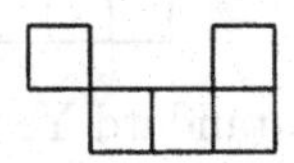

Then, according to the definition of $\to$ on $\mathbb{Z} \times \mathbb{Z}$, the natural labeling of S is taken row by row, from top left to bottom right as follows:

1.			2.
	3.	4.	5.

4.7 Definition. Let S be a shape and $n := |S|$. Any permutation $\pi \in \mathcal{S}_n$ is a *standard Young tableau of shape* S if the bijection

$$\alpha := \iota_S^{-1}\pi : (S, \leq_S) \longrightarrow (\underline{n}, \leq),$$

or, equivalently, the bijection

$$\sigma := \iota_S^{-1}\pi\iota_S : (S, \leq_S) \longrightarrow (S, \rightarrow_S)$$

is monotone. The set of all standard Young tableaux of shape S is denoted by SYT^S. We put

$$\mathrm{Z}^S := \sum_{\pi \in \mathsf{SYT}^S} \pi \in K\mathcal{S}_n \,.$$

4.8 Special case. $\mathsf{SYT}^{S(\pi)} = \{\pi\}$ and $\mathrm{Z}^{S(\pi)} = \pi$, for all permutations π.

Proof. Let $\pi \in \mathcal{S}_n$ and recall that, by definition, $i \leq_\pi j$ is equivalent to $i\pi \leq j\pi$, for all $i, j \in \underline{n}$. It follows that $\nu \in \mathsf{SYT}^{S(\pi)}$ if and only if $i\pi \leq j\pi$ implies $i\nu \leq j\nu$, for all $i, j \in \underline{n}$, which proves our claim. □

4.9 Example. To illustrate the notion of a standard Young tableau in the case of a subset $S \subseteq \mathbb{Z} \times \mathbb{Z}$ of order n, let $\pi \in \mathcal{S}_n$ and put $\alpha := \iota_S^{-1}\pi$. Following the conventions used for the components of a matrix, the image $x\alpha \in \mathbb{N}$ may be written into the cell x, for all $x \in S$. In other words, the images $1\pi, 2\pi, \ldots, n\pi$ are entered in the shape S, row-wise from top to bottom according to the natural labeling of S. Due to the illustration of the partial order $\leq_{\mathbb{Z}\times\mathbb{Z}}$, we then have $\pi \in \mathsf{SYT}^S$ if and only if the entries $y\alpha$ in all cells $y \in S$ weakly to the right of and above the cell $x \in S$ are larger then the entry $x\alpha$ in x. For example, for the shape

$S \simeq$

□			□
	□	□	□

mentioned in Example 4.6, the standard Young tableaux of shape S correspond to the bijections $S \rightarrow \underline{5}$ visualised as follows:

4			5
	1	2	3

3			5
	1	2	4

2			5
	1	3	4

1			5
	2	3	4

Accordingly, $\mathsf{SYT}^S = \{45123,\ 35124,\ 25134,\ 15234\}$.

4.10 Notation and Remarks. Let $(S, \rightarrow_S, \leq_S)$ be a shape and assume that $x, y \in S$ are incomparable with respect to $\leq_S$. Then we may define a shape

$$S(x,y) := (S, \rightarrow_S, \leq_{S(x,y)})$$

by

$$a \leq_{S(x,y)} b :\iff a \leq_S b \text{ or } (a \leq_S x \text{ and } y \leq_S b),$$

for all $a, b \in S$. The order $\leq_{S(x,y)}$ is the smallest refinement of $\leq_S$ such that $x \leq_{S(x,y)} y$. In particular, $\mathsf{SYT}^{S(x,y)} \subseteq \mathsf{SYT}^S$. Note that, repeating this procedure shows that $\mathsf{SYT}^T \neq \emptyset$ for all shapes T, by 4.8.

The following observation, which will be of interest throughout the next chapter, is due to Gessel [Ges84].

4.11 Lemma. *Let $(S, \rightarrow_S, \leq_S)$ be a shape and assume that $x, y \in S$ are incomparable with respect to $\leq_S$, then*

$$\mathrm{Z}^S = \mathrm{Z}^{S(x,y)} + \mathrm{Z}^{S(y,x)}.$$

Proof. It suffices to prove that SYT^S is the disjoint union of $\mathsf{SYT}^{S(x,y)}$ and $\mathsf{SYT}^{S(y,x)}$. As was mentioned already in 4.10, we have $\mathsf{SYT}^{S(x,y)} \subseteq \mathsf{SYT}^S$ and $\mathsf{SYT}^{S(y,x)} \subseteq \mathsf{SYT}^S$.

Assume that there exists an element $\pi \in \mathsf{SYT}^{S(x,y)} \cap \mathsf{SYT}^{S(y,x)}$, then $\alpha := \iota_S^{-1}\pi$ is monotone with respect to both $\leq_{S(x,y)}$ and $\leq_{S(y,x)}$. In particular, $x\alpha \leq y\alpha$ and $y\alpha \leq x\alpha$, hence $x\alpha = y\alpha$, contradicting injectivity of α.

Furthermore, for any $\pi \in \mathsf{SYT}^S$, the mapping $\alpha := \iota_S^{-1}\pi$ is injective again, and hence $x\alpha < y\alpha$, or $y\alpha < x\alpha$. This implies $\pi \in \mathsf{SYT}^{S(x,y)}$, or $\pi \in \mathsf{SYT}^{S(y,x)}$. □

For the sake of a better motivation, here is a typical application of 4.11 due to Stanley [Sta72], only the trivial part of which will be needed in what follows.

4.12 Corollary. *Let $(S, \rightarrow_S, \leq_S)$ be a shape of order n. If $\tilde{S} = (\underline{n}, \leq, \preceq_S)$ denotes the corresponding shape isomorphic to S as defined in 4.2, then*

$$\preceq_S = \bigcap_{\pi \in \mathsf{SYT}^S} \leq_\pi .$$

In particular, any shape T is isomorphic to S if and only if $\mathsf{SYT}^S = \mathsf{SYT}^T$, or, equivalently, $\mathrm{Z}^S = \mathrm{Z}^T$.

Proof. First observe that if $\preceq_S$ is a total order on $\underline{n}_|$, then $\preceq_S = \leq_\pi$ for the unique element π of SYT^S as asserted, by 4.8.

In general, for all $\pi \in \mathsf{SYT}^S$ and $i, j \in \underline{n}_|$ such that $i \preceq_S j$, we have $i\pi \leq j\pi$, hence

$$\preceq_S \subseteq \bigcap_{\pi \in \mathsf{SYT}^S} \leq_\pi .$$

The remaining inclusion will be proven by induction on the number of pairs of incomparable elements in $(S, \leq_S)$. Let $x, y \in S$ be incomparable with respect to $\leq_S$. Let $i, j \in \underline{n}_|$ be such that $i \leq_\pi j$ for all

$$\pi \in \mathsf{SYT}^S = \mathsf{SYT}^{S(x,y)} \cup \mathsf{SYT}^{S(y,x)},$$

then, inductively, $i \preceq_{S(x,y)} j$ and $i \preceq_{S(y,x)} j$. A quick look at the definition of $S(x, y)$ reveals that this implies $i \preceq_S j$, proving the second asserted inclusion, and completing the proof of the first claim.

If T is a second shape, then $S \simeq T$ certainly implies that any standard Young tableau of shape S is also a standard Young tableau of shape T and vice versa. Assume conversely that $\mathsf{SYT}^S = \mathsf{SYT}^T$, then

$$\preceq_S = \bigcap_{\pi \in \mathsf{SYT}^S} \leq_\pi = \bigcap_{\pi \in \mathsf{SYT}^T} \leq_\pi = \preceq_T$$

and thus $T \simeq (\underline{n}_|, \leq, \preceq_T) = (\underline{n}_|, \leq, \preceq_S) \simeq S$. □

4.13 Definition and Remark. Let $(S, \rightarrow_S, \leq_S)$ and $(T, \rightarrow_T, \leq_T)$ be shapes, then the shape $(U, \rightarrow_U, \leq_U)$ is a *semi-direct union* of S with T if there is a shape $(S', \rightarrow_{S'}, \leq_{S'})$ isomorphic to S, and a shape $(T', \rightarrow_{T'}, \leq_{T'})$ isomorphic to T such that $S' \cap T' = \emptyset$ and

$$U = S' \cup T', \quad \leq_U = \leq_{S'} \cup \leq_{T'} \text{ and } \rightarrow_U = \rightarrow_{S'} \cup \rightarrow_{T'} \cup (S' \times T').$$

The semi-direct nature of this union arises from the definition of the total order $\rightarrow_U$. For arbitrary S and T, such a semi-direct union U exists, by standard set theoretic arguments, and any two semi-direct unions of S with T are isomorphic. Furthermore, if $\tilde{S}$ and $\tilde{T}$ are shapes isomorphic to S and T, respectively, then any semi-direct union of $\tilde{S}$ with $\tilde{T}$ is isomorphic to U.

4.14 Example. Let S, T be shapes in $\mathbb{Z} \times \mathbb{Z}$ (with orders inherited from $\rightarrow$ and $\leq_{\mathbb{Z}\times\mathbb{Z}}$) belonging to the isomorphism classes of

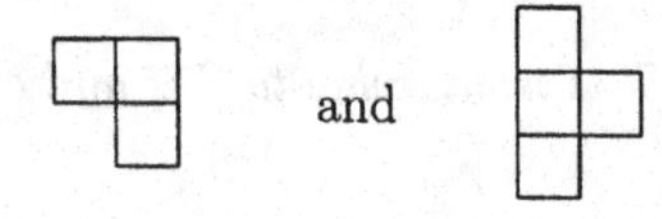

respectively, then any semi-direct union of S with T belongs to the isomorphism class of

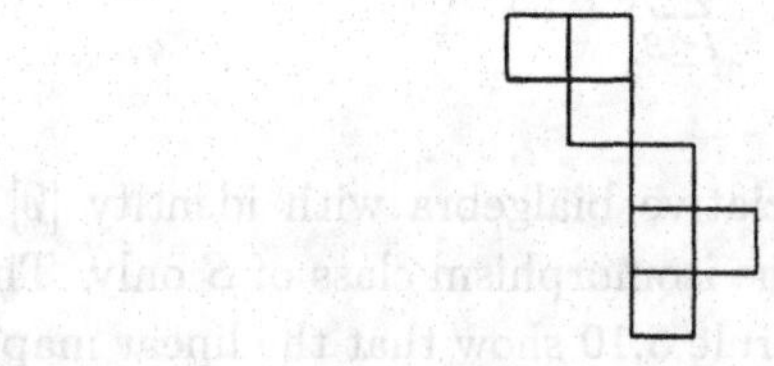

4.15 Definition. Let S be a shape. For any subset $T \subseteq S$, there are a total order and a partial order on T, obtained by restriction:

$$\rightarrow_T := \rightarrow_S \cap (T \times T) \quad \text{and} \quad \leq_T := \leq_S \cap (T \times T).$$

The shape $(T, \rightarrow_T, \leq_T)$ is a *sub-shape* of S.

Any sub-shape I of S is an *ideal* of S if the two conditions $y \in I$ and $x \leq_S y$ imply that $x \in I$, for all $x, y \in S$. We write $I \trianglelefteq S$ in this case.

4.16 Special case. Let $n \in \mathbb{N}$ and $\pi \in \mathcal{S}_n$, then the ideals of the π-shape $S(\pi)$ are given by

$$\emptyset,\ \underline{1}\pi^{-1},\ \underline{2}\pi^{-1},\ \ldots,\ \underline{n}\pi^{-1}.$$

Proof. Since $\leq_\pi$ is a total order on $\underline{n}$, the only ideals of $S(\pi) = (\underline{n}, \leq, \leq_\pi)$ are $I_k = \{\, i \in \underline{n} \,|\, i \leq_\pi k \,\}$, where $k \in \underline{n} \cup \{0\}$. The definition of $\leq_\pi$ implies that $i \leq_\pi k$ if and only if $i \in \underline{k\pi}\pi^{-1}$, hence $I_k = \underline{k\pi}\pi^{-1}$. □

4.17 Example. Let $\pi = 3\,4\,1\,5\,2 \in \mathcal{S}_5$, written as a word, then $3 <_\pi 5 <_\pi 1 <_\pi 2 <_\pi 4$ and $\pi^{-1} = 3\,5\,1\,2\,4$. The ideals I of $S(\pi)$ are given by

$$\emptyset, \quad \{3\}, \quad \{3,5\}, \quad \{3,5,1\}, \quad \{3,5,1,2\} \quad \text{and} \quad \underline{5}.$$

4.18 Remark. It is interesting here to consider a vector space V with K-basis the isomorphism classes of shapes (although this will not be of importance in what follows). The remarks given in 4.13 allow us to define the structure of an algebra on V by

$$[S] \cdot [T] = [U]$$

for all shapes S and T and bilinearity, where U is any semi-direct union of S with T. Here $[R]$ denotes the isomorphism class of R for any shape R.

Furthermore, it is clear that an isomorphism of shapes $\iota : S \to S'$ yields a one-to-one correspondence of the ideal lattices of S and S', by restriction.

Therefore, a coproduct Δ on V may be defined by

$$[S]\Delta = \sum_{I \trianglelefteq S} [I] \otimes [S \backslash I]$$

for all shapes S, and linearity.

In fact, $(V, \cdot, \Delta)$ is an associative bialgebra with identity $[\emptyset]$. Recall from 4.12 that Z^S depends on the isomorphism class of S only. The induction rule 5.5 and the restriction rule 5.10 show that the linear map, defined by

$$[S] \longmapsto \mathrm{Z}^S$$

for all shapes S, is an epimorphism of bialgebras from V onto the bialgebra $\mathcal{P}$ of permutations, which is studied in the next chapter.

Chapter 5

The Bialgebra $\mathcal{P}$ of Permutations

In this chapter a product, a coproduct, and a bilinear form on the direct sum

$$\mathcal{P} := \bigoplus_{n \in \mathbb{N}_0} K\mathcal{S}_n$$

are defined, which turn it into a self-dual bialgebra. This *bialgebra of permutations* was introduced by Malvenuto and Reutenauer in [MR95] and will be the general framework for the noncommutative character theory.

Here the approach is based on Stanley's theory of P-partitions discussed in the preceding chapter in terms of shapes and corresponding standard Young tableaux.

5.1 Definition. For all $n, m \in \mathbb{N}_0$, $\sigma \in \mathcal{S}_n$ and $\tau \in \mathcal{S}_m$, 4.12 and 4.13 show that

$$\sigma \star \tau := \mathrm{Z}^U$$

is well-defined, where U is an arbitrary semi-direct union of the σ-shape $S(\sigma)$ with the τ-shape $S(\tau)$. Bilinearity gives a product $\star$ on $\mathcal{P}$ called the convolution or *outer product* on $\mathcal{P}$. The definition of the semi-direct union then implies that

$$K\mathcal{S}_n \star K\mathcal{S}_m \subseteq K\mathcal{S}_{n+m}$$

for all $n, m \in \mathbb{N}_0$.

Recall that, for $k, l \in \mathbb{N}_0$, there is the transversal $\mathcal{S}^{k.l}$ of the right cosets of the Young subgroup $\mathcal{S}_{k.l} = \mathcal{S}_k \# \mathcal{S}_l$ in $\mathcal{S}_{k+l}$, consisting of all permutations $\nu \in \mathcal{S}_{k+l}$ which are increasing on $\underline{k}_|$ and on $\underline{k+l}_| \backslash \underline{k}_|$. Writing $\Xi^{k.l}$ for the sum of $\mathcal{S}^{k.l}$ in $K\mathcal{S}_{k+l}$ as in the introduction, allows us to give the following more explicit description of the outer product.

5.2 Proposition. *$\alpha \star \beta = (\alpha\#\beta)\Xi^{k.l}$, for all $k, l \in \mathbb{N}_0$, $\alpha \in K\mathcal{S}_k$ and $\beta \in K\mathcal{S}_l$.*

Proof. Let $k, l \in \mathbb{N}_0$ and set $n := k + l$. By linearity, it suffices to consider $\sigma \in \mathcal{S}_k$ instead of α and $\tau \in \mathcal{S}_l$ instead of β. A semi-direct union of $S(\sigma)$ with $S(\tau)$ can be constructed as follows. Let $S' := S(\sigma)$ be the σ-shape itself and define $\gamma : \underline{n-k} \to \underline{n} \setminus \underline{k}$, $i \mapsto i + k$, then, for the shape

$$T' := (\underline{n}\setminus\underline{k}, \to_{T'}, \leq_{T'}),$$

defined by $i \to_{T'} j :\iff i \leq j$ and $i \leq_{T'} j :\iff i\gamma^{-1}\tau \leq j\gamma^{-1}\tau$ for all $i, j \in \underline{n} \setminus \underline{k}$, we have $S' \cap T' = \emptyset$. Furthermore, the mapping $\gamma : S(\tau) \to T'$ is an isomorphism of shapes. Hence, for the semi-direct union U of the σ-shape with the τ-shape obtained from S' and T' as described in 4.13,

$$\sigma \star \tau = \sum_{\pi \in \mathsf{SYT}^U} \pi.$$

By definition, $\pi \in \mathsf{SYT}^U$ if and only if $i\sigma < j\sigma$ implies $i\pi < j\pi$ for all $i, j \in \underline{k}$, and $(i - k)\tau < (j - k)\tau$ implies $i\pi < j\pi$ for all $i, j \in \underline{n}\setminus\underline{k}$. Equivalently, $i(\sigma\#\tau) < j(\sigma\#\tau)$ implies $i\pi < j\pi$ whenever $i, j \in \underline{k}$ or $i, j \in \underline{n}\setminus\underline{k}$.

In other words, $\pi \in \mathsf{SYT}^U$ if and only if $\nu := (\sigma\#\tau)^{-1}\pi \in \mathcal{S}^{k.l}$. The proof is complete upon noting that the map $\nu \mapsto (\sigma\#\tau)\nu$ from $\mathcal{S}^{k.l}$ to $\mathcal{S}_n$ is injective, since $\mathcal{S}^{k.l}$ is a transversal of the right cosets of $\mathcal{S}_{k.l}$ in $\mathcal{S}_n$. □

5.3 Example. Let $k = 3$, $l = 2$, $\sigma = 2\,3\,1 \in \mathcal{S}_k$ and $\tau = 2\,1 \in \mathcal{S}_l$ written as words. Then $\sigma\#\tau = 2\,3\,1\,5\,4 \in \mathcal{S}_5$. Applying 5.2 yields

$$\begin{aligned}
\sigma \star \tau &= (\sigma\#\tau)\,\Xi^{3.2} \\
&= 23154(12345 + 12435 + 12534 + 13425 + 13524 \\
&\qquad + 14523 + 23415 + 23514 + 24513 + 34512) \\
&= \quad 23154 + 24153 + 25143 + 34152 + 35142 \\
&\qquad + 45132 + 34251 + 35241 + 45231 + 45321.
\end{aligned}$$

5.4 Remark. Recall from 2.9 that $(K\mathbb{N}^*, ., \delta)$ is a bialgebra, where . is the linear extension of the concatenation product on $\mathbb{N}^*$ and δ is defined by $n\delta = n \otimes \varnothing + \varnothing \otimes n$, for all $n \in \mathbb{N}$. As a general fact, this bialgebra structure gives rise to a *convolution product* $*$ on the ring $\operatorname{End} K\mathbb{N}^*$ of linear endomorphisms of $K\mathbb{N}^*$, as follows: If $f, g \in \operatorname{End} K\mathbb{N}^*$ and $u \in K\mathbb{N}^*$, then

$$u(f * g) = (u\delta)(f \otimes g)\mu_. .$$

The outer product $\star$ on $\mathcal{P}$ arises from this convolution product on $\operatorname{End} K\mathbb{N}^*$ once $\mathcal{P}$ is embedded into $\operatorname{End} K\mathbb{N}^*$, via *Polya action*:

$$\pi \mapsto f_\pi$$

for all permutations π. Here, for $\pi \in \mathcal{S}_k$, the linear map $f_\pi : K\mathbb{N}^* \to K\mathbb{N}^*$ is defined by sending $w = w_1 . \dots . w_n \in \mathbb{N}^*$ to

$$\pi w := w_{1\pi} . \dots . w_{n\pi}$$

or zero according as $k = n$ or not, for all $w \in \mathbb{N}^*$. Indeed, if $w \in \mathbb{N}^*$ has length n and $\pi \in \mathcal{S}_k$, $\sigma \in \mathcal{S}_l$, then

$$w(f_\pi * f_\sigma) = \sum_{J \subseteq \underline{n}} (w_{_J} f_\pi).(w_{_{CJ}} f_\sigma)$$

vanishes unless $n = k + l$. In this case,

$$w(f_\pi * f_\sigma) = \sum_{\substack{J \subseteq \underline{n} \\ |J| = k}} (\pi w_{_J}).(\sigma w_{_{CJ}}) = (\pi \# \sigma) \sum_{\nu \in \mathcal{S}^{k.l}} \nu w = w f_{\pi \star \sigma} ,$$

by 5.2. This shows $f_\pi * f_\sigma = f_{\pi \star \sigma}$. For more details, see [Reu93].

5.5 Induction Rule. *Let S, T be shapes, and let U be a semi-direct union of S with T, then*

$$Z^S \star Z^T = Z^U .$$

Proof. By 4.12, it may be assumed that $S \cap T = \emptyset$, $U = S \cup T$ and that the orders of U arise directly from those of S and T as described in 4.13. Let n and m be the number of pairs (x, y) in $S \times S$ and $T \times T$, respectively, such that x and y are incomparable with respect to $\leq_S$ and $\leq_T$, respectively. The formula is proved by induction on $n + m$. If $n + m = 0$, then $\leq_S$ and $\leq_T$ are total orders on S and T, hence there are permutations σ and τ such that $S \simeq S(\sigma)$ and $T \simeq S(\tau)$, by 4.3. Applying 4.12, 4.13, 4.8 and the definition 5.1 yields

$$Z^S \star Z^T = Z^{S(\sigma)} \star Z^{S(\tau)} = \sigma \star \tau = Z^U.$$

Let $n+m > 0$. Then there exist elements $x, y \in S$, or $x, y \in T$ incomparable with respect to $\leq_S$, or $\leq_T$. Assume that $x, y \in S$. The other case is proved along the same lines. Observe that $U_1 := U(x,y)$ and $U_2 := U(y,x)$ are semi-direct unions of $S_1 := S(x,y)$ and $S_2 := S(y,x)$, respectively, with T. Hence, by 4.11 and induction, it follows that

$$Z^S \star Z^T = (Z^{S_1} + Z^{S_2}) \star Z^T = Z^{S_1} \star Z^T + Z^{S_2} \star Z^T = Z^{U_1} + Z^{U_2} = Z^U$$

as asserted. □

5.6 Theorem. *$(\mathcal{P}, \star)$ is an associative algebra with identity* $\mathrm{id}_0 = \emptyset$*, the unique element of* $\mathcal{S}_0$.

Proof. The fact that $\emptyset \in \mathcal{S}_0$ acts as the identity in $(\mathcal{P}, \star)$ is immediate from 5.2, while the associativity of $\star$ may be derived easily from the associativity of the semi-direct union: It suffices to show that $(\sigma\star\pi)\star\tau = \sigma\star(\pi\star\tau)$, for all permutations σ, π, τ. Let R, S, T be pairwise disjoint shapes such that $R \simeq S(\sigma)$, $S \simeq S(\pi)$, and $T \simeq S(\tau)$. Then the shapes $V = R \cup S$, $W = S \cup T$, and $U = V \cup T$ with the respective order relations defined in 4.13, are semi-direct unions of $S(\sigma)$ with $S(\pi)$, of $S(\pi)$ with $S(\tau)$, and of V with $S(\tau)$. Simply by definition, U is also a semi-direct union of $S(\sigma)$ with W. Hence, applying 4.12 and 5.5 a number of times, we obtain

$$\begin{aligned}(\sigma \star \pi) \star \tau &= (Z^R \star Z^S) \star Z^T \\ &= Z^V \star Z^T \\ &= Z^U \\ &= Z^R \star Z^W \\ &= Z^R \star (Z^S \star Z^T) \\ &= \sigma \star (\pi \star \tau).\end{aligned}$$

□

5.7 Definition and Remark. The coproduct $\downarrow$ on $\mathcal{P}$ is defined by

$$\pi\!\downarrow\; := \sum_{I \trianglelefteq S(\pi)} Z^I \otimes Z^{S(\pi)\setminus I}$$

for all $n \in \mathbb{N}_0$, $\pi \in \mathcal{S}_n$, and linearity. Note that

$$K\mathcal{S}_n\downarrow \subseteq \sum_{k=0}^{n} K\mathcal{S}_k \otimes K\mathcal{S}_{n-k} ,$$

for all $n \in \mathbb{N}_0$.

Observing that $(\mathcal{S}^{k.l})^{-1}$ is a transversal of the left cosets of $\mathcal{S}_{k.l}$ in $\mathcal{S}_{k+l}$, for all $k, l \in \mathbb{N}_0$, allows us to give the following explicit description of $\pi\downarrow$.

5.8 Proposition. *Let $n \in \mathbb{N}$ and $\pi \in \mathcal{S}_n$. For each $k \in \underline{n} \cup \{0\}$, choose $\nu_k \in \mathcal{S}^{k.(n-k)}$ and $\alpha_k \in \mathcal{S}_k$, $\beta_k \in \mathcal{S}_{n-k}$ so that $\pi = \nu_k^{-1}(\alpha_k \# \beta_k)$, then*

$$\pi\downarrow = \sum_{k=0}^{n} \alpha_k \otimes \beta_k .$$

Viewed as a word, the permutation α_k is obtained by lining up the elements $1, 2 \ldots, k$ from π in their present order. Similarly, the permutation β_k is obtained by reading out the numbers $k+1, \ldots, n$ from π from left to right and subtracting k from each of them.

Proof. Recall that $I := \underline{k}\pi^{-1}$ is the unique ideal of order k in $S(\pi)$, by 4.16. Furthermore, SYT^I and $\mathsf{SYT}^{S(\pi)\setminus I}$ are singletons, by 4.3 and 4.8, hence $\alpha_k \in \mathsf{SYT}^I$ and $\beta_k \in \mathsf{SYT}^{S(\pi)\setminus I}$ remain to be shown.

For all $i, j \in \underline{k}$, $i < j$ implies

$$i\alpha_k\pi^{-1} = i(\alpha_k \# \beta_k)\pi^{-1} = i\,\nu_k < j\,\nu_k = j(\alpha_k \# \beta_k)\pi^{-1} = j\alpha_k\pi^{-1},$$

hence $\alpha_k\pi^{-1} : \underline{k} \to I$ respects the usual order on $\underline{k}$ and I. It follows that $\alpha_k\pi^{-1} = \iota_I$ and $\alpha_k = \iota_I\pi \in \mathsf{SYT}^I$.

$\beta_k \in \mathsf{SYT}^{S(\pi)\setminus I}$ is shown analogously. □

5.9 Examples. As in 4.17, let $\pi = 3\,4\,1\,5\,2 \in \mathcal{S}_5$, written as a word, then

$$\pi\downarrow = \emptyset \otimes 3\,4\,1\,5\,2 + 1 \otimes 2\,3\,4\,1 + 1\,2 \otimes 1\,2\,3 + 3\,1\,2 \otimes 1\,2 + 3\,4\,1\,2 \otimes 1 + 3\,4\,1\,5\,2 \otimes \emptyset .$$

In particular, $(\mathcal{P}, \downarrow)$ is not cocommutative. For all $n \in \mathbb{N}_0$, 5.8 implies

$$\mathsf{id}_n\downarrow = \sum_{k=0}^{n} \mathsf{id}_k \otimes \mathsf{id}_{n-k} .$$

For the order reversing involution ρ_n in $\mathcal{S}_n$, defined by $i\rho_n := n - i + 1$ for all $i \in \underline{n}$, that is,

$$\rho_n = n\,(n-1)\,(n-2) \cdots 2\,1$$

when written as a word, there is the analogous rule

$$\rho_n\downarrow = \sum_{k=0}^{n} \rho_k \otimes \rho_{n-k} .$$

It is left to the reader to show that, more generally, if $q = q_1.\dots.q_k \models n$ and $\rho^q := \rho_{q_1}\#\cdots\#\rho_{q_k} \in \mathcal{S}_n$, then

$$\rho^q\downarrow = \sum_{\substack{r,s\in\mathbb{N}^* \\ r.s=q}} \rho^r \otimes \rho^s + \sum_{\substack{r,s\in\mathbb{N}^*\setminus\{\varnothing\} \\ r\sqcup s=q}} \rho^r \otimes \rho^s,$$

where

$$r\sqcup s := r_1.\dots.r_{k-1}.(r_k+s_1).s_2.\dots.s_l$$

for all $r = r_1.\dots.r_k, s = s_1.\dots.s_l \in \mathbb{N}^*\setminus\{\varnothing\}$.

5.10 Restriction Rule. $Z^S\downarrow = \sum_{I\trianglelefteq S} Z^I \otimes Z^{S\setminus I}$, *for any shape* S.

Proof. Assume first that $\leq_S$ is a total order on S, then there is a permutation π such that $S \simeq S(\pi)$, by 4.3. From 4.12, 4.8 and the definition of the coproduct $\downarrow$,

$$Z^S\downarrow = Z^{S(\sigma)}\downarrow = \sigma\downarrow = \sum_{I\trianglelefteq S(\pi)} Z^I \otimes Z^{S(\pi)\setminus I} = \sum_{I\trianglelefteq S} Z^I \otimes Z^{S\setminus I}.$$

We proceed by induction on the number of pairs of incomparable elements in $(S, \leq_S)$. Let $x, y \in S$ be incomparable with respect to $\leq_S$. Concerning the ideal structure of $S(x,y)$, observe, for any $J \subseteq S$, that:

(i) If J is an ideal of $S(x,y)$, then $y \in J$ implies that $x \in J$, as $x \leq_{S(x,y)} y$.

(ii) If $x, y \in J$, then J is an ideal of $S(x,y)$ if and only if there exists an ideal I of S such that $x, y \in I$ and $J = I(x,y)$. In this case, the shapes $S(x,y) \setminus J$ and $S \setminus I$ coincide.

(iii) If $x, y \in S \setminus J$, then J is an ideal of $S(x,y)$ if and only if J is an ideal of S. In this case, we have $S(x,y) \setminus J = (S \setminus I)(x,y)$.

(iv) If $x \in J$ and $y \in S \setminus J$, then J is an ideal of $S(x,y)$ if and only if J is an ideal of S. In this case, the shapes $S(x,y) \setminus J$ and $S \setminus I$ again coincide.

Using these observations for both $S(x,y)$ and $S(y,x)$, it follows by induction that

$$
\begin{aligned}
Z^S\downarrow &= Z^{S(x,y)}\downarrow + Z^{S(y,x)}\downarrow, \qquad \text{by 4.11}\\
&= \sum_{J \trianglelefteq S(x,y)} Z^J \otimes Z^{S(x,y)\setminus J} + \sum_{J \trianglelefteq S(y,x)} Z^J \otimes Z^{S(y,x)\setminus J}\\
&= \sum_{\substack{J \trianglelefteq S(x,y)\\ x,y\in J}} Z^J \otimes Z^{S(x,y)\setminus J}\\
&\qquad + \sum_{\substack{J \trianglelefteq S(x,y)\\ x,y\in S\setminus J}} Z^J \otimes Z^{S(x,y)\setminus J} + \sum_{\substack{J \trianglelefteq S(x,y)\\ x\in J,\, y\in S\setminus J}} Z^J \otimes Z^{S(x,y)\setminus J}\\
&\quad + \sum_{\substack{J \trianglelefteq S(y,x)\\ x,y\in J}} Z^J \otimes Z^{S(y,x)\setminus J}\\
&\qquad + \sum_{\substack{J \trianglelefteq S(y,x)\\ x,y\in S\setminus J}} Z^J \otimes Z^{S(y,x)\setminus J} + \sum_{\substack{J \trianglelefteq S(y,x)\\ y\in J,\, x\in S\setminus J}} Z^J \otimes Z^{S(y,x)\setminus J}, \text{ by (i)}\\
&= \sum_{\substack{I \trianglelefteq S\\ x,y\in I}} (Z^{I(x,y)} + Z^{I(y,x)}) \otimes Z^{S\setminus I}\\
&\qquad + \sum_{\substack{I \trianglelefteq S\\ x,y\in S\setminus I}} Z^I \otimes (Z^{(S\setminus I)(x,y)} + Z^{(S\setminus I)(y,x)})\\
&\qquad\qquad + \sum_{\substack{I \trianglelefteq S\\ |\{x,y\}\cap I|=1}} Z^I \otimes Z^{S\setminus I}, \quad \text{by (ii), (iii), (iv)}\\
&= \sum_{\substack{I \trianglelefteq S\\ x,y\in I}} Z^I \otimes Z^{S\setminus I} + \sum_{\substack{I \trianglelefteq S\\ x,y\in S\setminus I}} Z^I \otimes Z^{S\setminus I} + \sum_{\substack{I \trianglelefteq S\\ |\{x,y\}\cap I|=1}} Z^I \otimes Z^{S\setminus I},\\
&\hspace{20em} \text{by 4.11}\\
&= \sum_{I \trianglelefteq S} Z^I \otimes Z^{S\setminus I},
\end{aligned}
$$

which completes the proof. □

5.11 Theorem. *$(\mathcal{P}, \star, \downarrow)$ is a bialgebra.*

Proof. It suffices to show $(\sigma \star \tau)\downarrow = \sigma\downarrow \star_\otimes \tau\downarrow$ for all permutations σ, τ. Let U be a semi-direct union of $S(\sigma)$ with $S(\tau)$, say $U = S \cup T$ such that $S \cap T = \emptyset$, $S \simeq S(\sigma)$ and $T \simeq S(\tau)$. Then $I \subseteq U$ is an ideal of U if and only if there exists an ideal I_1 of S, and an ideal I_2 of T such that $I = I_1 \cup I_2$. In this case, I is a semi-direct union of I_1 with I_2, while $U \backslash I = (S\backslash I_1) \cup (T \backslash I_2)$ is a semi-direct union of $(S\backslash I_1)$ with $(T\backslash I_2)$.

Using 5.10 and 5.5, the asserted homomorphism rule is derived as follows:

$$\begin{aligned}
(\sigma \star \tau)\downarrow = \mathrm{Z}^U\downarrow &= \sum_{I \trianglelefteq U} \mathrm{Z}^I \otimes \mathrm{Z}^{U\backslash I} \\
&= \sum_{I_1 \trianglelefteq S} \sum_{I_2 \trianglelefteq T} \mathrm{Z}^{I_1 \cup I_2} \otimes \mathrm{Z}^{(S\backslash I_1)\cup(T\backslash I_2)} \\
&= \sum_{I_1 \trianglelefteq S} \sum_{I_2 \trianglelefteq T} \mathrm{Z}^{I_1} \star \mathrm{Z}^{I_2} \otimes \mathrm{Z}^{S\backslash I_1} \star \mathrm{Z}^{T\backslash I_2} \\
&= \sum_{I_1 \trianglelefteq S} \sum_{I_2 \trianglelefteq T} (\mathrm{Z}^{I_1} \otimes \mathrm{Z}^{S\backslash I_1}) \star_\otimes (\mathrm{Z}^{I_2} \otimes \mathrm{Z}^{T\backslash I_2}) \\
&= \mathrm{Z}^S\downarrow \star_\otimes \mathrm{Z}^T\downarrow \\
&= \sigma\downarrow \star_\otimes \tau\downarrow\,.
\end{aligned}$$

□

5.12 Definition and Remark. Define a bilinear form $(\,\cdot\,,\,\cdot\,)_\mathcal{P}$ on $\mathcal{P}$ by

$$(\sigma, \tau)_\mathcal{P} := \begin{cases} 1 & \text{if } \sigma = \tau^{-1} \\ 0 & \text{otherwise} \end{cases}$$

for all permutations σ, τ. This form is regular and symmetric. Furthermore, $K\mathcal{S}_n \perp K\mathcal{S}_m$ for all $n, m \in \mathbb{N}_0$ such that $n \neq m$.

5.13 Notation. Denote by $(\,\cdot\,,\,\cdot\,)_{\mathcal{P}\otimes\mathcal{P}}$ the unique bilinear form on $\mathcal{P} \otimes \mathcal{P}$ inherited from $(\,\cdot\,,\,\cdot\,)_\mathcal{P}$ as described in 2.6, then

$$(\alpha_1 \otimes \alpha_2, \beta_1 \otimes \beta_2)_{\mathcal{P}\otimes\mathcal{P}} = (\alpha_1, \beta_1)_\mathcal{P} (\alpha_2, \beta_2)_\mathcal{P}$$

for all $\alpha_1, \alpha_2, \beta_1, \beta_2 \in \mathcal{P}$.

This chapter concludes with the important

5.14 Reciprocity Law. *The bialgebra $(\mathcal{P}, \star, \downarrow)$ is self-dual with respect to $(\cdot, \cdot)_{\mathcal{P}}$, that is, for all $\alpha, \beta, \gamma \in \mathcal{P}$,*

$$(\alpha \star \beta, \gamma)_{\mathcal{P}} = (\alpha \otimes \beta, \gamma\downarrow)_{\mathcal{P}\otimes\mathcal{P}} .$$

Proof. Let $a, b, n \in \mathbb{N}_0$ and $\sigma \in \mathcal{S}_a$, $\tau \in \mathcal{S}_b$, $\pi \in \mathcal{S}_n$. Using bilinearity, it suffices to show that

$$(\sigma \star \tau, \pi)_{\mathcal{P}} = (\sigma \otimes \tau, \pi\downarrow)_{\mathcal{P}\otimes\mathcal{P}} .$$

Both terms are zero unless $a + b = n$, by 5.1 and 5.7. Let $a + b = n$, then the scalar product $(\sigma \star \tau, \pi)_{\mathcal{P}}$ is one or zero according as $\pi^{-1} = (\sigma \# \tau)\nu$ for some $\nu \in \mathcal{S}^{a.b}$ or not, by 5.2.

For all $k \in \underline{n} \cup \{0\}$, choose $\nu_k \in \mathcal{S}^{k.(n-k)}$, $\alpha_k \in \mathcal{S}_k$ and $\beta_k \in \mathcal{S}_{n-k}$ such that $\pi = \nu_k^{-1}(\alpha_k \# \beta_k)$, then the scalar product

$$(\sigma \otimes \tau, \pi\downarrow)_{\mathcal{P}\otimes\mathcal{P}} = \sum_{k=0}^{n} (\sigma \otimes \tau, \alpha_k \otimes \beta_k)_{\mathcal{P}\otimes\mathcal{P}} = (\sigma \otimes \tau, \alpha_a \otimes \beta_a)_{\mathcal{P}\otimes\mathcal{P}}$$

is one or zero according as $\sigma^{-1} = \alpha_a$ and $\tau^{-1} = \beta_a$ or not, by 5.8. But this condition is equivalent to $\pi = \nu_a^{-1}(\alpha_a \# \beta_a) = \nu_a^{-1}(\sigma^{-1} \# \tau^{-1})$. The proof is complete. □

Chapter 6

Frames

Alfred Young was the first to use the notion of a *frame*, that is, a *convex* shape contained in $\mathbb{Z} \times \mathbb{Z}$, in the representation theory of the symmetric group (see, for example, [You28; You34]). The aim of this chapter is to analyse two of the major notions introduced in Chapter 5 — the induction rule 5.5 and the restriction rule 5.10 — in the special case of frames. For the time being, this may be viewed as an illustration of those parts of the abstract theory introduced so far. The applications will be seen in Part III.

As a byproduct, there is a sub-bialgebra of $\mathcal{P}$, the Jöllenbeck algebra $\mathcal{F}$, which is linearly generated by the elements Z^F (F a frame). It was this bialgebra upon which the noncommutative theory introduced in [Jöl98] was based.

The direct sum $\mathcal{D}$ of the Solomon descent algebras $\mathcal{D}_n$, $n \in \mathbb{N}_0$, is contained in $\mathcal{F}$. In fact, Solomon's noncommutative Young characters Ξ^q are among the linear generators Z^F of $\mathcal{F}$, as a proper choice of frames F will show at the end of this chapter. In this way, we see that the bialgebra $(\mathcal{D}, \star, \downarrow)$ considered in the introduction is a sub-bialgebra of the algebra $\mathcal{P}$ of permutations.

Recall that any subset U of a partially-ordered set $(M, \preceq)$ is *convex* if, for all $x, z \in U$ and $y \in M$, $x \preceq y \preceq z$ implies that $y \in U$.

6.1 Definition. Let F be a finite subset of $\mathbb{Z} \times \mathbb{Z}$. The total order on F inherited from the total order $\rightarrow$ on $\mathbb{Z} \times \mathbb{Z}$ is denoted by $\rightarrow_F$, while the partial order on F inherited from $\leq_{\mathbb{Z}\times\mathbb{Z}}$ is denoted by $\leq_F$. The shape $(F, \rightarrow_F, \leq_F)$ is a *frame* if F is a convex subset of $(\mathbb{Z} \times \mathbb{Z}, \leq_{\mathbb{Z}\times\mathbb{Z}})$.

6.2 Example. Up to isomorphism, the frames of order 3 may be illustrated as follows:

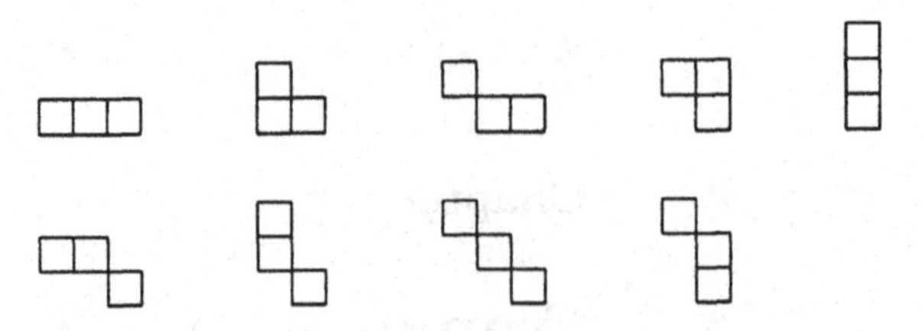

6.3 Definition. Let F be a frame, then $F^j := (\mathbb{Z} \times \{j\}) \cap F$ is the j-*th column* and $F_i := (\{i\} \times \mathbb{Z}) \cap F$ is the i-*th row* of F, for all $i, j \in \mathbb{Z}$.

Convexity allows a first simplification by the observation that, in the case of a frame F, the standard Young tableaux of shape F may be characterised locally as follows.

6.4 Proposition. *Let F be a frame, $n := |F|$, $\pi \in \mathcal{S}_n$ and $\alpha := \iota_F^{-1}\pi$, then π is a standard Young tableau of shape F if and only if α is increasing in the rows (from left to right) and decreasing in the columns (from top to bottom) of F, that is, if*

$$(i,j)\alpha \leq (i,j+1)\alpha \quad \text{and} \quad (k,l)\alpha \leq (k+1,l)\alpha$$

for all $i, j, k, l \in \mathbb{Z}$ such that $(i,j), (i,j+1), (k,l), (k+1,l) \in F$.

As F is convex, this is readily seen from

$$(i,j) \leq_{\mathbb{Z}\times\mathbb{Z}} (i,j+1) \leq_{\mathbb{Z}\times\mathbb{Z}} \cdots \leq_{\mathbb{Z}\times\mathbb{Z}} (i,l) \leq_{\mathbb{Z}\times\mathbb{Z}} (i+1,l) \leq_{\mathbb{Z}\times\mathbb{Z}} \cdots \leq_{\mathbb{Z}\times\mathbb{Z}} (k,l)$$

for all $(i,j), (k,l) \in \mathbb{Z} \times \mathbb{Z}$ such that $(i,j) \leq_{\mathbb{Z}\times\mathbb{Z}} (k,l)$.

An alternative and possibly more common description of frames may be given in terms of *partition frames.*

6.5 Definition. For any partition $p \in \mathbb{N}^*$, the set

$$\mathsf{F}(p) := \bigcup_{i=1}^{\ell(p)} \{i\} \times \underline{p_i}$$

is called the *partition frame* corresponding to p. Let q be a second partition. Set

$$\mathsf{F}(p\backslash q) := \mathsf{F}(p)\backslash\mathsf{F}(q)$$

and, for brevity, $\mathsf{Z}^p := \mathsf{Z}^{\mathsf{F}(p)}$ and $\mathsf{Z}^{p\backslash q} := \mathsf{Z}^{\mathsf{F}(p)\backslash\mathsf{F}(q)}$.

6.6 Proposition. *For all partitions p, q, $\mathsf{F}(p\backslash q)$ is a frame. Conversely, for any frame F, there exist partitions $p, q \in \mathbb{N}^*$ such that $F \simeq \mathsf{F}(p\backslash q)$ and, in addition, $\mathsf{F}(q) \subseteq \mathsf{F}(p)$.*

Proof. Let $x, y, z \in \mathbb{Z} \times \mathbb{Z}$ such that $x \leq_{\mathbb{Z}\times\mathbb{Z}} y \leq_{\mathbb{Z}\times\mathbb{Z}} z$ and $x, z \in \mathsf{F}(p\backslash q)$, then $x \notin \mathsf{F}(q)$ implies that $y \notin \mathsf{F}(q)$, while $z \in \mathsf{F}(p)$ implies that $y \in \mathsf{F}(p)$, hence $y \in \mathsf{F}(p\backslash q)$. It follows that $\mathsf{F}(p\backslash q)$ is convex as asserted.

The second part may be proved by induction on the order of F. However, the straightforward but somewhat tedious line of reasoning is left to the reader. □

Which description of frames is preferred depends on the application and, of course, on personal taste.

6.7 Notation. Denote the linear subspace of $\mathcal{P}$ generated by the elements Z^F, F a frame, by $\mathcal{F}$. Then

$$\mathcal{F} = \bigoplus_{n\in\mathbb{N}_0} \mathcal{F}_n\,,$$

where $\mathcal{F}_n$ is the linear span of the elements Z^F such that F is a frame of order n, for all $n \in \mathbb{N}_0$.

The following propositions, which show that $\mathcal{F}$ is a sub-bialgebra of $\mathcal{P}$, yield helpful illustrations of the results derived in Chapter 5. To begin with, *products* in $\mathcal{F}$ are analysed by constructing a semi-direct union of two frames. This demonstrates the function of 4.11 and 5.5.

6.8 Definition and Remark. Let F and G be frames. If $F = \emptyset$ or $G = \emptyset$, let

$$U := F \cup G,$$

while, for $F \neq \emptyset$ and $G \neq \emptyset$, denote the largest element of $(F, \rightarrow_F)$ by x and the smallest element of $(G, \rightarrow_G)$ by z and put

$$G' := G - z + x + (-1, 1)$$

and

$$U := F \cup G'.$$

Then U is a frame, and a semi-direct union of F with G, called the *coupling* of F with G.

If F and G are nonempty, the situation may be illustrated as follows. Start with

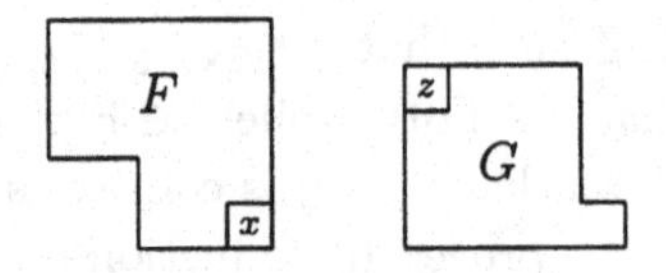

in arbitrary relative positions, then

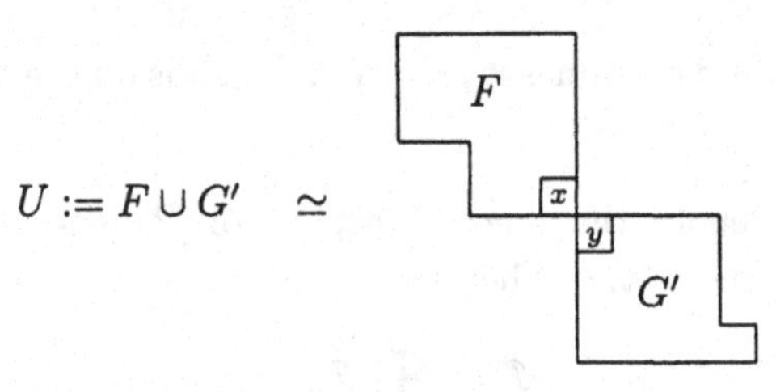

where $y := x + (-1, 1)$ is the smallest element of $(G', \rightarrow)$.

Proof. There is nothing to prove in the case where $F = \emptyset$ or $G = \emptyset$.

Let F and G be nonempty, then the choice of x and z means that the frames F and G' are disjoint. Furthermore, by 4.5, G' and G are isomorphic. Let $x = (a, b)$. The cell $y := (a - 1, b + 1)$ is minimal in G' with respect to $\rightarrow$. Convexity of F and maximality of x in $(F, \rightarrow)$ imply $j \leq b$ for all $(i, j) \in F$. Similarly, $b + 1 \leq l$ for all $(k, l) \in G'$, as y is minimal in G' and G' is also convex. Therefore $j < l$, for all $(i, j) \in F$ and $(k, l) \in G'$. Furthermore,

$$(i, j) \rightarrow x \rightarrow y \rightarrow (k, l)$$

yields $i \geq k$. Hence (i, j) and (k, l) are incomparable with respect to $\leq_{\mathbb{Z}\times\mathbb{Z}}$. Now it is readily seen that U is also convex, and indeed a semi-direct union of F with G. □

6.9 Proposition. *Let F and G be nonempty frames, and let U be the coupling of F with G, then*

$$\mathrm{Z}^F \star \mathrm{Z}^G = \mathrm{Z}^U.$$

Furthermore, if F and G are nonempty and G' is defined as in 6.8, then

$$\mathrm{Z}^U = \mathrm{Z}^{F\cup(G'+(1,0))} + \mathrm{Z}^{F\cup(G'-(0,1))}.$$

Proof. The first equality is immediate from 5.5 and 6.8.

Let F and G be nonempty and denote by x the largest element of $(F, \rightarrow_F)$, then $y = x + (-1, 1)$ is the smallest element of $(G', \rightarrow_{G'})$ and x,

y are incomparable in $(U, \leq_U)$. Furthermore,

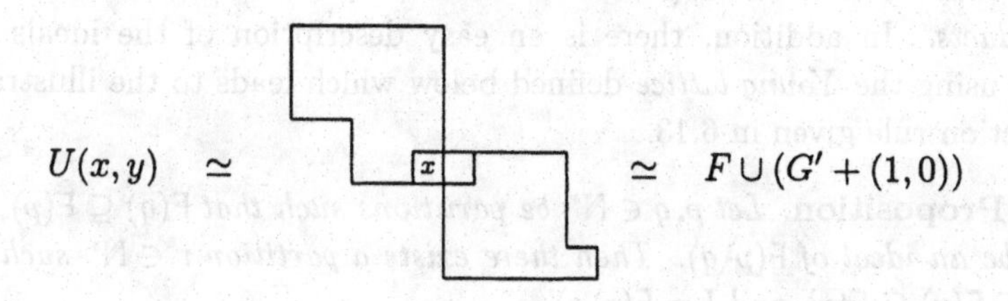

and

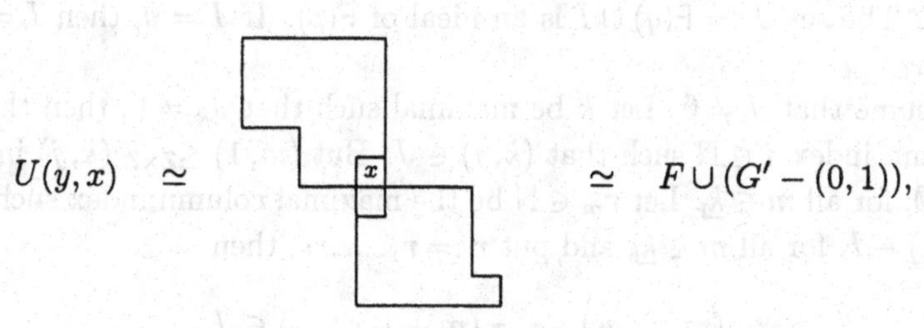

as is easily seen. The second equality thus follows from 4.11. □

6.10 Example. Let $k, m \in \mathbb{N}$ and consider the special case in which F is a vertical bar of order k, and G is a horizontal bar of order m,

$$F \simeq \mathsf{F}(1^k) \simeq \cdots, \qquad G \simeq \mathsf{F}(m) \simeq \cdots,$$

where $1^k := 1.\ldots.1$ (k factors). The resulting formula

$$\mathrm{Z}^{1^k} \star \mathrm{Z}^m = \mathrm{Z}^{m.1^k} + \mathrm{Z}^{(m+1).1^{k-1}},$$

pictorially

$$x,\ y \;=\; x \;+\; x$$

is useful. For example, it readily implies the linear relation

$$\sum_{k=0}^{n} (-1)^k \mathrm{Z}^{1^k} \star \mathrm{Z}^{n-k} = 0$$

for all $n \in \mathbb{N}$, where, by definition, $\mathrm{Z}^0 = \emptyset \in \mathcal{S}_0$ is the identity of $\mathcal{P}$.

If F is a frame and I is an ideal of F, then the definition of an ideal implies that both I and $F\backslash I$ are convex. Hence, by 5.10, $\mathcal{F}$ is closed under *coproducts.* In addition, there is an easy description of the ideals of a frame using the *Young lattice* defined below which leads to the illustrative restriction rule given in 6.13.

6.11 Proposition. *Let $p, q \in \mathbb{N}^*$ be partitions such that $\mathsf{F}(q) \subseteq \mathsf{F}(p)$, and let I be an ideal of $\mathsf{F}(p\backslash q)$. Then there exists a partition $r \in \mathbb{N}^*$ such that $\mathsf{F}(q) \subseteq \mathsf{F}(r) \subseteq \mathsf{F}(p)$ and $I = \mathsf{F}(r\backslash q)$.*

Proof. The set $J := \mathsf{F}(q) \cup I$ is an ideal of $\mathsf{F}(p)$. If $J = \emptyset$, then $I = \emptyset = \mathsf{F}(q\backslash q)$.

Assume that $J \neq \emptyset$. Let k be maximal such that $J_k \neq \emptyset$, then there is a column index $j \in \mathbb{N}$ such that $(k, j) \in J$. But $(m, 1) \leq_{\mathbb{Z}\times\mathbb{Z}} (k, j)$ implies $J_m \neq \emptyset$, for all $m \in \underline{k}_{\mathsf{l}}$. Let $r_m \in \mathbb{N}$ be the maximal column index such that $(m, r_m) \in I$, for all $m \in \underline{k}_{\mathsf{l}}$, and put $r := r_1. \ldots .r_k$, then

$$(m, r_{m+1}) \leq_{\mathbb{Z}\times\mathbb{Z}} (m+1, r_{m+1}) \in J$$

implies that $(m, r_{m+1}) \in J$ and so $r_{m+1} \leq r_m$, for all $m \in \underline{k-1}_{\mathsf{l}}$. Hence r is a partition, and $J = \mathsf{F}(r)$. The claim follows. □

6.12 Definition. Define a partial order $\subseteq$ on the set of partitions in $\mathbb{N}^*$ by

$$q \subseteq p :\Longleftrightarrow \mathsf{F}(q) \subseteq \mathsf{F}(p).$$

This partially-ordered set is a lattice, the *Young lattice.*

Combining 6.11 with 5.10, yields:

6.13 Corollary. *Let $p, q \in \mathbb{N}^*$ be partitions such that $q \subseteq p$, then*

$$\mathsf{Z}^{p\backslash q}\downarrow = \sum_r \mathsf{Z}^{r\backslash q} \otimes \mathsf{Z}^{p\backslash r},$$

where the sum is taken over all partitions r such that $q \subseteq r \subseteq p$.

Now 6.9 and 6.13 imply indeed:

6.14 Corollary. *$\mathcal{F}$ is a sub-bialgebra of $(\mathcal{P}, \star, \downarrow)$, called the* Jöllenbeck algebra.

Here is the link to Solomon's theory.

6.15 Definition and Remark. Let $q = q_1.\dots.q_l \in \mathbb{N}^*$ and denote by $s_i := q_1 + \cdots + q_i$ the i-th partial sum of q, for all $i \in \underline{l}$, then the words $p := s_l.s_{l-1}.\dots.s_1$ and $r := s_{l-1}.\dots.s_1$ are partitions. The frame

$$\mathsf{HS}(q) := \mathsf{F}(p) \backslash \mathsf{F}(r)$$

may be visualised by

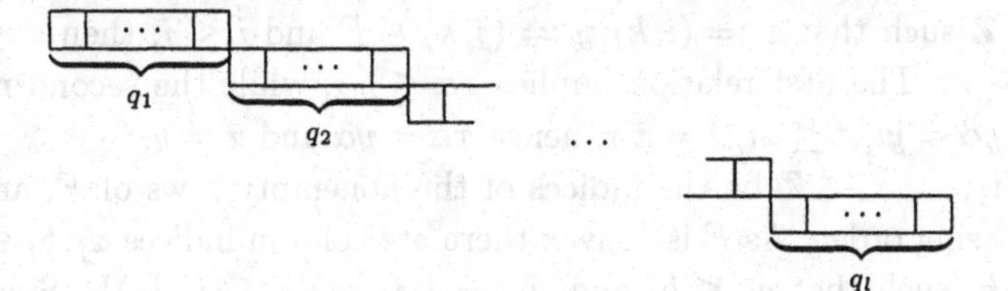

Any frame isomorphic to $\mathsf{HS}(q)$ is called a *horizontal strip of type q*.

Recall the definition of P^q given in the introduction to observe that $\pi \in \mathsf{SYT}^{\mathsf{HS}(q)}$ if and only if π is increasing on the successive blocks $P_1^q, P_2^q, \dots, P_l^q$ in $\underline{n}$ of order $q_1, q_2, \dots, q_l$, that is,

$$\mathcal{S}^q = \mathsf{SYT}^{\mathsf{HS}(q)} \quad \text{and} \quad \Xi^q = \mathsf{Z}^{\mathsf{HS}(q)}.$$

6.16 Corollary. *The linear span $\mathcal{D}$ of the elements $\Xi^q = \mathsf{Z}^{\mathsf{HS}(q)}$, $q \in \mathbb{N}^*$, is a sub-bialgebra of $\mathcal{F}$. Furthermore,*

$$\Xi^r \star \Xi^q = \Xi^{r.q}$$

for all $q, r \in \mathbb{N}^*$ and

$$\Xi^n \downarrow = \sum_{k=0}^{n} \Xi^k \otimes \Xi^{n-k}$$

for all $n \in \mathbb{N}$, where $\Xi^0 := \emptyset \in \mathcal{S}_0$ is the identity of $(\mathcal{P}, \star)$.

Proof. The product rule is a special case of 6.9, since $\mathsf{HS}(r.q)$ is isomorphic to the coupling of $\mathsf{HS}(r)$ with $\mathsf{HS}(q)$. Furthermore, $\Xi^n = \mathsf{id}_n$ for all $n \in \mathbb{N}$, hence the coproduct rule was given in 5.9 already. □

For later use, this chapter concludes with a characterisation of horizontal strips.

6.17 Proposition. *Let F be a frame of order n, then F is a horizontal strip if and only if $\mathsf{id}_n \in \mathsf{SYT}^F$.*

Proof. Let $q \models n$ such that $F \simeq \mathsf{HS}(q)$, then

$$\mathsf{id}_n \in \mathcal{S}^q = \mathsf{SYT}^{\mathsf{HS}(q)} = \mathsf{SYT}^F.$$

Conversely, assume that $\mathsf{id}_n \in \mathsf{SYT}^F$, so that

$$\alpha := \iota_F^{-1}\mathsf{id}_n = \iota_F^{-1} : (F, \leq_F) \to (\underline{n}, \leq)$$

is monotone. Then each column of F contains at most one cell. Indeed, if $i, j, k \in \mathbb{Z}$ such that $x := (i, k), y := (j, k) \in F$ and $i \leq j$, then $x \leq_{\mathbb{Z}\times\mathbb{Z}} y$ and $y \to x$. The first relation implies $x\alpha \leq y\alpha$, while the second relation implies $y\alpha = y\iota_F^{-1} \leq x\iota_F^{-1} = x\alpha$, hence $x\alpha = y\alpha$ and $x = y$.

Let $i_1, \ldots, i_n \in \mathbb{Z}$ be the indices of the nonempty rows of F, arranged in decreasing order. As F is convex there are column indices $a_j, b_j \in \mathbb{Z}$, for all $j \in \underline{k}$, such that $a_j \leq b_j$ and $F_{i_j} = \{(i_j, a_j), \ldots, (i_j, b_j)\}$. Since each column of F contains at most one cell, it follows that $b_j < a_{j+1}$ for all $j \in \underline{k-1}$, by the convexity of F, and thus

$$F \simeq \mathsf{HS}((b_1 - a_1 + 1).\cdots.(b_k - a_k + 1)).$$

□

Chapter 7

Epimorphisms

In this chapter, the crucial link is given between the noncommutative algebra $\mathcal{P}$ of permutations and the commutative algebra $\mathcal{C}$ of class functions.

Throughout, Π_n is a primitive element in $K\mathcal{S}_n$ for all $n \in \mathbb{N}$, in the following sense.

7.1 Definition and Remarks. Let $n \in \mathbb{N}$. An element $\varphi \in \mathcal{P}$ is *primitive* if $\varphi\!\downarrow\, = \varphi \otimes \emptyset + \emptyset \otimes \varphi$. The set of all primitive elements in $\mathcal{P}$ is a linear subspace of $\mathcal{P}$, since the coproduct $\downarrow$ is linear.

For each $q = q_1.\dots.q_k \in \mathbb{N}^*$, put $\Pi_q := \Pi_{q_1} \star \cdots \star \Pi_{q_k}$, then

$$\begin{aligned}\Pi_q\!\downarrow\, &= (\Pi_{q_1}\!\downarrow) \star_\otimes \cdots \star_\otimes (\Pi_{q_k}\!\downarrow)\\ &= (\Pi_{q_1} \otimes \emptyset + \emptyset \otimes \Pi_{q_1}) \star_\otimes \cdots \star_\otimes (\Pi_{q_k} \otimes \emptyset + \emptyset \otimes \Pi_{q_k})\\ &= \sum_{J \subseteq \underline{k}} \Pi_{q_J} \otimes \Pi_{q_{\complement J}},\end{aligned}$$

since $(\mathcal{P}, \star, \downarrow)$ is a bialgebra.

7.2 Example. Let $n \in \mathbb{N}$. The element

$$\Pi_n := \sum_{q \models n} (-1)^{n-\ell(q)} \rho^q \in K\mathcal{S}_n$$

is primitive, since the coproduct rule for the elements ρ^q mentioned in 5.9 implies

$$\Pi_n\!\downarrow\, = \Pi_n \otimes \emptyset + \emptyset \otimes \Pi_n + \sum_{\substack{r,s \in \mathbb{N}^* \setminus \{\varnothing\} \\ r.s \models n}} \left((-1)^{n-\ell(r.s)} + (-1)^{n-\ell(r \sqcup s)}\right) \rho^r \otimes \rho^s,$$

and $\ell(r.s) = \ell(r \sqcup s) + 1$ for all $r, s \in \mathbb{N}^* \backslash \{\varnothing\}$.

Note that $(\Pi_n, \mathrm{id}_n)_{\mathcal{P}} = (\Pi_n, \rho^{1^n})_{\mathcal{P}} = 1$ and $(\Pi_n, \Pi_n)_{\mathcal{P}} = 2^{n-1}$ for all $n \in \mathbb{N}$, since $(\rho^q)^{-1} = \rho^q$ for all $q \in \mathbb{N}^*$.

Linear bases of the space of all primitive elements in $\mathcal{P}$ have been constructed in [AS; DHT02].

7.3 Definition and Remarks. Define a linear mapping $c_\Pi : \mathcal{P} \to \mathcal{C}$ by

$$c_\Pi(\varphi)(C_p) := (\varphi, \Pi_p)_{\mathcal{P}}$$

for all $n \in \mathbb{N}$, $\varphi \in K\mathcal{S}_n$ and $p \vdash n$. In other words, the value $c_\Pi(\varphi)(\pi)$ of the c_Π-image of $\varphi \in K\mathcal{S}_n$ on any element $\pi \in C_p$ is $(\varphi, \Pi_p)_{\mathcal{P}}$.

Note that c_Π is *graded*, that is, it maps $K\mathcal{S}_n$ into $\mathcal{C}\ell_K(\mathcal{S}_n)$ for all $n \in \mathbb{N}_0$. Furthermore, it is convenient to observe that, if $\gamma = \sum_{m \in \mathbb{N}_0} \gamma_m \in \mathcal{P}$ with $\gamma_m \in K\mathcal{S}_m$ for all $m \in \mathbb{N}_0$ and p is a partition of n, then

$$\begin{aligned} c_\Pi(\gamma)(C_p) &= \sum_{m \in \mathbb{N}_0} c_\Pi(\gamma_m)(C_p) \\ &= c_\Pi(\gamma_n)(C_p) \\ &= (\gamma_n, \Pi_p)_{\mathcal{P}} \\ &= \sum_{m \in \mathbb{N}_0} (\gamma_m, \Pi_p)_{\mathcal{P}} \\ &= (\gamma, \Pi_p)_{\mathcal{P}}. \end{aligned}$$

7.4 Theorem. *$c_\Pi : (\mathcal{P}, \star) \to (\mathcal{C}, \bullet)$ is a homomorphism of algebras.*

Proof. Let $\alpha, \beta \in \mathcal{P}$ and $p = p_1. \ldots .p_l \in \mathbb{N}^*$ be a partition. Using self-duality of $\mathcal{P}$ and $\mathcal{C}$ and applying 3.6 and 3.13, we obtain

$$\begin{aligned} c_\Pi(\alpha \star \beta)(C_p) &= (\alpha \star \beta, \Pi_p)_{\mathcal{P}} \\ &= (\alpha \otimes \beta, \Pi_p\!\downarrow)_{\mathcal{P} \otimes \mathcal{P}} \\ &= \sum_{J \subseteq \underline{l}} (\alpha, \Pi_{p_J})_{\mathcal{P}} (\beta, \Pi_{p_{CJ}})_{\mathcal{P}} \\ &= \sum_{J \subseteq \underline{l}} c_\Pi(\alpha)(C_{p_J})\, c_\Pi(\beta)(C_{p_{CJ}}) \end{aligned}$$

$$= \sum_{J \subseteq \underline{l}} (c_\Pi(\alpha), \mathrm{ch}_{p_J})_{\mathcal{C}} (c_\Pi(\beta), \mathrm{ch}_{p_{CJ}})_{\mathcal{C}}$$

$$= (c_\Pi(\alpha) \otimes c_\Pi(\beta), \mathrm{ch}_p\!\downarrow)_{\mathcal{C}\otimes\mathcal{C}}$$

$$= (c_\Pi(\alpha) \bullet c_\Pi(\beta), \mathrm{ch}_p)_{\mathcal{C}}$$

$$= (c_\Pi(\alpha) \bullet c_\Pi(\beta))(C_p),$$

since p_J and p_{CJ} are partitions, for all $J \subseteq \underline{l}$. □

7.5 Example. For the element $\Xi^n = \mathrm{id}_n \in \mathcal{S}_n$,

$$c_\Pi(\Xi^n)(C_p) = (\Xi^n, \Pi_p)_{\mathcal{P}} = (\mathrm{id}_n, \Pi_p)_{\mathcal{P}}$$

is the coefficient of id_n in Π_p, for all $p \vdash n$. Let $p = p_1. \dots .p_l$ and set $\tilde{p} = p_1. \dots .p_{l-1}$, then the coproduct rule for id_n given in 5.9 and self-duality of $\mathcal{P}$ yield

$$c_\Pi(\Xi^n)(C_p) = (\mathrm{id}_n, \Pi_{\tilde{p}} \star \Pi_{p_l})_{\mathcal{P}}$$

$$= \sum_{k=0}^{n} (\mathrm{id}_k \otimes \mathrm{id}_{n-k}, \Pi_{\tilde{p}} \otimes \Pi_{p_l})_{\mathcal{P}\otimes\mathcal{P}}$$

$$= (\mathrm{id}_{n-p_l}, \Pi_{\tilde{p}})_{\mathcal{P}} (\mathrm{id}_{p_l}, \Pi_{p_l})_{\mathcal{P}}$$

$$= \cdots$$

$$= (\mathrm{id}_{p_1}, \Pi_{p_1})_{\mathcal{P}} \cdots (\mathrm{id}_{p_l}, \Pi_{p_l})_{\mathcal{P}},$$

by a simple induction on $l = \ell(p)$. If, in particular, $(\mathrm{id}_n, \Pi_n)_{\mathcal{P}} = 1$ for all $n \in \mathbb{N}$, then $c_\Pi(\Xi^n)$ is the trivial character ξ^n of $\mathcal{S}_n$.

Combining this example with Theorem 7.4, yields:

7.6 Corollary. *Assume that* $(\Pi_n, \mathrm{id}_n)_{\mathcal{P}} = 1$ *for all* $n \in \mathbb{N}$, *then the restriction of* c_Π *to* $\mathcal{D}_n$ *coincides with Solomon's epimorphism* c_n, *for all* $n \in \mathbb{N}$.

In particular, in this case, c_Π *is onto.*

Indeed, by 6.16, 7.4 and 3.2,

$$\begin{aligned} c_\Pi(\Xi^q) &= c_\Pi(\Xi^{q_1} \star \cdots \star \Xi^{q_k}) \\ &= c_\Pi(\Xi^{q_1}) \bullet \cdots \bullet c_\Pi(\Xi^{q_k}) \\ &= \xi^{q_1} \bullet \cdots \bullet \xi^{q_k} \\ &= \xi^q \end{aligned}$$

for all $q = q_1 . \dots . q_k \in \mathbb{N}^*$.

Another consequence of the above theorem concerns the images of the elements Π_q and concludes this chapter.

7.7 Corollary. *Set* $a_n := (\Pi_n, \Pi_n)_{\mathcal{P}}/n$ *for all* $n \in \mathbb{N}$, *then*

$$c_\Pi(\Pi_q) = a_{q_1} \cdots a_{q_k} \mathsf{ch}_q$$

for all $q = q_1 . \dots . q_k \in \mathbb{N}^*$.

Proof. Primitivity of Π_n and self-duality of $\mathcal{P}$ imply $c_\Pi(\Pi_n)(C_p) = (\Pi_n, \Pi_p)_{\mathcal{P}} = 0$ for all partitions $p \neq n$ in $\mathbb{N}^*$. This shows $c_\Pi(\Pi_n) = a_n \mathsf{ch}_n$ and also

$$c_\Pi(\Pi_q) = c_\Pi(\Pi_{q_1} \star \cdots \star \Pi_{q_k}) = c_\Pi(\Pi_{q_1}) \bullet \cdots \bullet c_\Pi(\Pi_{q_k}) = a_{q_1} \cdots a_{q_k} \mathsf{ch}_q$$

for all $q = q_1 . \dots . q_k \in \mathbb{N}^*$, by 7.4 and 3.7. □

Chapter 8

The Coplactic Algebra $\mathcal{Q}$

The Robinson–Schensted correspondence stated below is the combinatorial core of noncommutative character theory.

A new non-algorithmic proof of the Robinson–Schensted correspondence is given in Appendix C. It is well adapted to the current setting and builds on the detailed analysis of Knuth's relations given in [BJ99]. For other proofs and general reference on the subject, see [Ful97; Knu73; Lee96], for example.

8.1 Definition and Remark. It is convenient for our purposes to denote the interval of positive integers with margins a and b by $\langle a, b\rangle$, for all a and b in $\mathbb{N}$, that is:

$$\langle a, b\rangle := \begin{cases} \{c \in \mathbb{N} \mid a \le c \le b\} & \text{if } a \le b, \\ \{c \in \mathbb{N} \mid b \le c \le a\} & \text{if } b < a. \end{cases}$$

Let $n \in \mathbb{N}_0$ and $\pi, \sigma \in \mathcal{S}_n$, then σ is a *plactic neighbour* of π if there exists an index $i \in \underline{n-1}$ such that $\sigma = \tau_{n,i}\pi$ and $(i-1)\pi$ or $(i+2)\pi$ is contained in $\langle i\pi, (i+1)\pi\rangle$. Here $\tau_{n,i} := (i\ (i+1))$ denotes the transposition in $\mathcal{S}_n$ swapping i and $i+1$, and $0\pi := 0$ or $(n+1)\pi := 0$ if necessary.

If π and σ are viewed as words, then this means that σ is obtained from π by swapping the i-th and $(i+1)$-th letter, and that the $(i-1)$-th letter to the left, or the $(i+1)$-th letter to the right of $i\pi\,(i+1)\pi$ in the image line of π, is in between of $i\pi$ and $(i+1)\pi$ with respect to the usual ordering of $\underline{n}$. This relation is symmetric.

The smallest equivalence $\sim$ on $\mathcal{S}_n$ containing the plactic neighbourhood is the *plactic equivalence*. The corresponding equivalence classes in $\mathcal{S}_n$ are the *plactic classes* in $\mathcal{S}_n$.

Let $\mathsf{SYT}^p := \mathsf{SYT}^{\mathsf{F}(p)}$ and $\mathsf{syt}^p = |\mathsf{SYT}^p|$ for all partitions p.

Robinson–Schensted correspondence. *Let $n \in \mathbb{N}$, then there is a bijection*

$$\mathcal{S}_n \longrightarrow \bigcup_{p \vdash n} \mathsf{SYT}^p \times \mathsf{SYT}^p, \quad \pi \longmapsto (P(\pi), Q(\pi))$$

such that for all $\pi, \sigma \in \mathcal{S}_n$

(i) $P(\pi^{-1}) = Q(\pi)$,

(ii) $P(\pi) = P(\sigma)$ *if and only if* $\pi \sim \sigma$,

(iii) $P(\pi) \sim \pi$.

Such a bijection was defined algorithmically by Schensted in [Sch61]. The first component P is the *Schensted P-symbol*, while Q is the *Schensted Q-symbol.* It is now folklore that a different construction given earlier by Robinson [Rob38] yields the same bijection. Both algorithms involve highly asymmetric constructions of the P- and the Q-symbol.

However, the two components are linked by (i), which is known as Schützenberger's theorem [Sch63]. By (ii), the equivalence on $\mathcal{S}_n$ arising from equality of P-symbols is the plactic equivalence defined above — a result of Knuth [Knu73]. These relations on the symmetric groups $\mathcal{S}_n$ can be extended to a congruence of the free monoid $\mathbb{N}^*$. The quotient monoid consisting of the corresponding congruence classes in $\mathbb{N}^*$ has a significant importance in algebraic combinatorics in general, due to the work of Lascoux and Schützenberger [LS81], who called it the *plactic monoid.*

The plactic relations stated in (iii) are readily derived from Schensted's algorithmic description of the P-symbol (see, for instance, [BJ99]) and allow one to give the following description of the P- and the Q-symbol.

8.2 Proposition. *Let $n \in \mathbb{N}$ and $\sigma \in \mathcal{S}_n$, then the P-symbol of σ is the* unique *standard Young tableau $\alpha \in \bigcup_{p \vdash n} \mathsf{SYT}^p$ contained in the plactic class of σ, while the Q-symbol of σ is the* unique *standard Young tableau $\beta \in \bigcup_{p \vdash n} \mathsf{SYT}^p$ contained in the plactic class of σ^{-1}.*

Proof. It suffices to prove the first part concerning the P-symbol, by Schützenberger's theorem.

The P-symbol of σ is contained in $\bigcup_{p \vdash n} \mathsf{SYT}^p$ and in the plactic class of σ, by (iii). To prove uniqueness, consider first an arbitrary standard Young tableau α in $\bigcup_{p \vdash n} \mathsf{SYT}^p$. Surjectivity of the Robinson–Schensted correspondence allows us to choose a permutation $\nu \in \mathcal{S}_n$ with P-symbol α.

Then ν and α are in plactic relation, by (iii), hence $P(\alpha) = P(\nu) = \alpha$, by Knuth's theorem (ii). Thus, if α and α' are two standard Young tableaux contained in the plactic class of σ, then $\alpha = P(\alpha) = P(\alpha') = \alpha'$, by (ii) again. □

8.3 Definition and Remark. Let $n \in \mathbb{N}$ and $\pi, \sigma \in \mathcal{S}_n$, then σ is a *coplactic neighbour* of π if σ^{-1} is a plactic neighbour of π^{-1}. Equivalently, there exists an index $i \in \underline{n-1}$ such that $\sigma = \pi\tau_{n,i}$ and $i-1$ or $i+2$ is contained in

$$\langle i, i+1 \rangle_\pi := \langle i\pi^{-1}, (i+1)\pi^{-1} \rangle \pi,$$

the segment in the image line of π bordered by and including i and $i+1$.

The reflexive and transitive cover of the coplactic neighbourhood is the *coplactic equivalence*, so that σ and π are in coplactic relation if and only if their inverses are in plactic relation. The equivalence classes in $\mathcal{S}_n$ arising from the coplactic equivalence are given by equality of Q-symbols, according to (i) and (ii), and referred to as *coplactic classes.* The number of coplactic classes in $\mathcal{S}_n$ is equal to the number of possible Q-symbols, that is: $|\bigcup_{p \vdash n} \mathsf{SYT}^p|$.

For each $n \in \mathbb{N}_0$, denote the linear span of all sums ΣA of coplactic classes A in $\mathcal{S}_n$ by $\mathcal{Q}_n$ and set

$$\mathcal{Q} := \bigoplus_{n \in \mathbb{N}_0} \mathcal{Q}_n .$$

Note that the elements ΣA, where A is a coplactic class in $\mathcal{S}_n$, constitute a linear basis of $\mathcal{Q}_n$, since two coplactic classes in $\mathcal{S}_n$ are equal or disjoint, so that

$$\dim \mathcal{Q}_n = \sum_{p \vdash n} \mathsf{syt}^p .$$

The linear space $\mathcal{Q}$ is a sub-bialgebra of the bialgebra $\mathcal{P}$ of permutations. This is the *coplactic algebra*, which was discovered by Poirier and Reutenauer [PR95] and which contains the Jöllenbeck algebra $\mathcal{F}$ defined in the previous chapter. The coplactic algebra may be viewed as an algebraisation of Schensted's combinatorial construction and turns out to be the natural framework for the noncommutative character theory.

Before the above mentioned properties of $\mathcal{Q}$ are derived, the Robinson–Schensted correspondence for $\mathcal{S}_4$ shall be illustrated as an example. Observe that, in general, the symmetric group $\mathcal{S}_n$ is subdivided into several cells indexed by partitions p of n, according to the shape of the P-symbol and of the Q-symbol of a permutation $\pi \in \mathcal{S}_n$.

8.4 Definition and Remark. Let $n \in \mathbb{N}$ and $p \vdash n$, then

$$\mathcal{G}^p := \{\pi \in \mathcal{S}_n \mid P(\pi) \in \mathsf{SYT}^p\} = \{\pi \in \mathcal{S}_n \mid Q(\pi) \in \mathsf{SYT}^p\}$$

is the *Greene cell* corresponding to p. The Greene cells are the equivalence classes arising from the smallest equivalence on $\mathcal{S}_n$ containing both the plactic and the coplactic equivalence. Each Greene cell is thus a union of plactic classes and a union of coplactic classes. Note that SYT^p is contained in $\mathcal{G}^p$, for all $p \vdash n$, by 8.2.

We mention that Greene [Gre74] discovered a way to determine the shape of the P-symbol of a permutation $\pi \in \mathcal{S}_n$ directly, that is, without recourse to Schensted's algorithm (see the final part of Appendix C).

8.5 Example. The Robinson–Schensted correspondence for the symmetric group $\mathcal{S}_4$ is now illustrated. The Greene cells $\mathcal{G}^4$, $\mathcal{G}^{3.1}$, $\mathcal{G}^{2.2}$, $\mathcal{G}^{2.1.1}$ and $\mathcal{G}^{1.1.1.1}$ in $\mathcal{S}_4$ are considered as coordinate systems. The plactic classes constitute the columns, and the coplactic classes the rows of each cell. Two plactic (respectively, coplactic) neighbours are connected by a vertical (respectively, horizontal) bar, which is labelled by the corresponding swap position.

For each $p \vdash 4$, the elements of SYT^p constitute the first row of the corresponding cell $\mathcal{G}^p$, while the elements of $(\mathsf{SYT}^p)^{-1}$ constitute its first column. We shall see later on that SYT^p is a coplactic class for any partition p (see 8.8). The intersection of SYT^p with $(\mathsf{SYT}^p)^{-1}$ (the upper left corner, or "origin" of $\mathcal{G}^p$) contains a unique element $\pi \in \mathsf{SYT}^p$, and $\pi = \pi^{-1}$. In fact, starting with these two axes of $\mathcal{G}^p$, it is possible to round out the cell $\mathcal{G}^p$ for arbitrary p using proper plactic and coplactic relations.

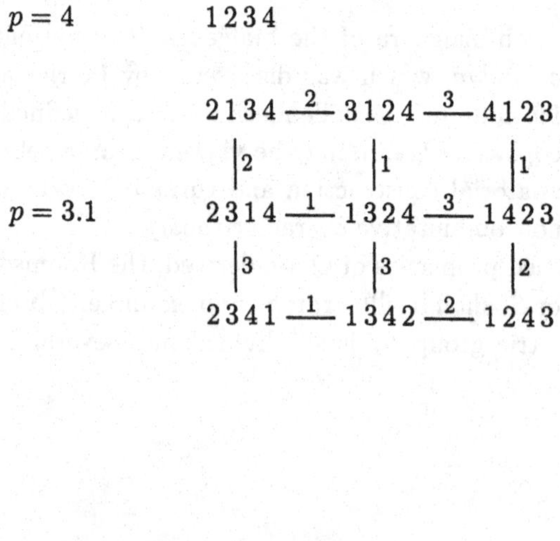

$p = 4$ 1234

$$\begin{array}{ccccc}
2134 & \overset{2}{\text{———}} & 3124 & \overset{3}{\text{———}} & 4123 \\
\Big|\,{\scriptstyle 2} & & \Big|\,{\scriptstyle 1} & & \Big|\,{\scriptstyle 1} \\
2314 & \overset{1}{\text{———}} & 1324 & \overset{3}{\text{———}} & 1423 \\
\Big|\,{\scriptstyle 3} & & \Big|\,{\scriptstyle 3} & & \Big|\,{\scriptstyle 2} \\
2341 & \overset{1}{\text{———}} & 1342 & \overset{2}{\text{———}} & 1243
\end{array}$$

$p = 3.1$

$$p = 2.2 \qquad \begin{array}{ccc} 3412 & \overset{2}{\text{———}} & 2413 \\ \big|{\scriptstyle 2} & & \big|{\scriptstyle 2} \\ 3142 & \overset{2}{\text{———}} & 2143 \end{array}$$

$$p = 2.1.1 \qquad \begin{array}{ccccc} 3214 & \overset{3}{\text{———}} & 4213 & \overset{2}{\text{———}} & 4312 \\ \big|{\scriptstyle 3} & & \big|{\scriptstyle 3} & & \big|{\scriptstyle 2} \\ 3241 & \overset{3}{\text{———}} & 4231 & \overset{1}{\text{———}} & 4132 \\ \big|{\scriptstyle 2} & & \big|{\scriptstyle 1} & & \big|{\scriptstyle 1} \\ 3421 & \overset{2}{\text{———}} & 2431 & \overset{1}{\text{———}} & 1432 \end{array}$$

$$p = 1.1.1.1 \qquad 4321$$

Once this is done, the Robinson–Schensted correspondence is described pictorially as follows. To compute the P- and the Q-symbol of $\pi = 2431$, say, first locate the cell ("coordinate system") containing π, that is: $\mathcal{G}^{2.1.1}$. Now move from π to the top edge of the cell, its "P-axis", and reach the standard Young tableau $\alpha = 4213$ of shape $p = 2.1.1$, the "P-coordinate" of π. Then move from π to the left edge of the cell, its "Q-axis", and reach the inverted standard Young tableau $\beta = 3421 \in (\mathsf{SYT}^p)^{-1}$, the "$Q$-coordinate" of π. This yields

$$(P(\pi), Q(\pi)) = (\alpha, \beta^{-1}) = \left(\begin{array}{|c|c} \cline{1-1} 4 & \\ \cline{1-1} 2 & \\ \hline 1 & \multicolumn{1}{c|}{3} \\ \hline \end{array} \; , \; \begin{array}{|c|c} \cline{1-1} 4 & \\ \cline{1-1} 3 & \\ \hline 1 & \multicolumn{1}{c|}{2} \\ \hline \end{array} \right),$$

by 8.2. Injectivity of the Robinson–Schensted correspondence guarantees that no permutation occurs in two different coordinate systems and that each horizontal and each vertical line in the same coordinate system intersect in a unique point.

Furthermore, in this illustration, Schützenberger's theorem says that taking inverses exchanges the coordinates. In particular, each permutation $\pi \in \mathcal{S}_n$ with $\pi = \pi^{-1}$ is located on the main diagonal of its cell, and each coplactic (and each plactic) class in $\mathcal{S}_n$ contains a unique such element.

Let it be mentioned that, as n grows large, the graph structures of the cells in $\mathcal{S}_n$ grow much more complicated.

We proceed by deriving the algebraic properties of the linear space $\mathcal{Q}$, step by step.

8.6 Theorem. *$\mathcal{Q}$ is a sub-bialgebra of $(\mathcal{P}, \star, \downarrow)$, called the* coplactic algebra.

Proof. Let $n, m \in \mathbb{N}_0$ and choose coplactic classes A in $\mathcal{S}_n$ and A' in $\mathcal{S}_m$ respectively.

There exists a subset C of $\mathcal{S}_{n+m}$ such that $\Sigma A \star \Sigma A' = \Sigma C$, by 5.2, since both the embedding $\# : \mathcal{S}_n \times \mathcal{S}_m \to \mathcal{S}_{n.m}$ and the product map $\mathcal{S}_{n.m} \times \mathcal{S}^{n.m} \to \mathcal{S}_{n+m}$ are injective. We need to show that C is a union of coplactic classes in order to prove that $\Sigma A \star \Sigma A' \in \mathcal{Q}$.

Let $\pi \in C$ and choose a coplactic neighbour σ of π in $\mathcal{S}_{n+m}$. There exist permutations $\alpha \in A$, $\alpha' \in A'$ and $\nu \in \mathcal{S}^{n.m}$ such that $\pi = (\alpha \# \alpha')\nu$, by 5.2. Furthermore, $\sigma = \pi\tau_{n+m,i}$ for some $i \in \underline{n+m}$ such that $i-1$ or $i+2$ is contained in $\langle i, i+1\rangle_\pi$.

If the intersection $\{i, i+1\} \cap \underline{n}\nu$ is a singleton, then $\nu\tau_{n+m,i} \in \mathcal{S}^{n.m}$, hence $\sigma = \pi\tau_{n+m,i} = (\alpha\#\alpha')(\nu\tau_{n+m,i}) \in C$.

Assume now that both $x = i\nu^{-1}$ and $y = (i+1)\nu^{-1}$ are contained in $\underline{n}$. Then $\nu \in \mathcal{S}^{n.m}$ implies $x < y$. In fact, it even follows that $y = x+1$. As a consequence, $\nu\tau_{n+m,i}\nu^{-1} = (i\nu^{-1}\ (i+1)\nu^{-1}) = \tau_{n+m,x}$ and thus

$$\sigma = \pi\tau_{n+m,i} = (\alpha\#\alpha')\nu\tau_{n+m,i} = (\alpha\#\alpha')\tau_{n+m,x}\nu = (\alpha\tau_{n,x}\#\alpha')\nu.$$

Furthermore, $i\pi^{-1} = x\alpha^{-1}$ and $(i+1)\pi^{-1} = (x+1)\alpha^{-1}$. If $i-1$ is contained in $\langle i, i+1\rangle_\pi$, then $(i-1)\pi^{-1} \in \underline{n}$, hence also $(i-1)\nu^{-1} \in \underline{n}$. So $\nu \in \mathcal{S}^{n.m}$ implies $(i-1)\nu^{-1} = x-1$. Similarly, if $i+2$ is contained in $\langle i, i+1\rangle_\pi$, then $(i+2)\pi^{-1} \in \underline{n}$ implies $(i+2)\nu^{-1} \in \underline{n}$ and $(i+2)\nu^{-1} = x+2$.

It follows that $(x-1)\alpha^{-1} = (i-1)\pi^{-1}$ or $(x+2)\alpha^{-1} = (i+2)\pi^{-1}$ is contained in the interval $\langle x\alpha^{-1}, (x+1)\alpha^{-1}\rangle$. In other words, $\alpha\tau_{n,x}$ is a coplactic neighbour of α, hence $\alpha\tau_{n,x} \in A$ and $\sigma \in C$.

The case where both i and $i+1$ are not contained in $\underline{n}\nu$ may be dealt with analogously: here we find that $\sigma = (\alpha\#\alpha'\tau_{m,x-n})\nu$ and that $\alpha'\tau_{m,x-n} \in A'$.

We have proven that $\mathcal{Q}$ is a subalgebra of $(\mathcal{P}, \star)$. In order to show it is also a sub-coalgebra, choose coefficients $c_{\alpha,\beta} \in K$ such that

$$\Sigma A\!\downarrow\; = \sum c_{\alpha,\beta}\, \alpha \otimes \beta,$$

where the sum ranges over all permutations α, β such that $\alpha \in \mathcal{S}_k$ and $\beta \in \mathcal{S}_{n-k}$ for some $k \in \underline{n} \cup \{0\}$.

We fix $\alpha \in \mathcal{S}_k$ and $\beta \in \mathcal{S}_{n-k}$ and prove that $c_{\alpha,\beta} = c_{\alpha',\beta'}$ whenever α is in coplactic relation with α' and β is in coplactic relation with β'.

In fact, it is enough to show that $c_{\alpha,\beta} = c_{\alpha',\beta}$ for each coplactic neighbour α' of α and $c_{\alpha,\beta} = c_{\alpha,\beta'}$ for each coplactic neighbour β' of β, for this implies the above statement and thus $\Sigma A\!\downarrow\, \in \mathcal{Q}\otimes\mathcal{Q}$. Observe first that 5.8 implies

$$c_{\alpha,\beta} = |A_{\alpha,\beta}|,$$

where $A_{\alpha,\beta}$ denotes the set of all $\pi \in A$ such that $\nu\pi = \alpha\#\beta$ for some $\nu \in \mathcal{S}^{k.(n-k)}$. Fix $\nu \in \mathcal{S}^{k.(n-k)}$ and choose a coplactic neighbour α' of α, say, $\alpha' = \alpha\tau_{k,x}$. Then the map $\pi \mapsto \pi' := \pi\tau_{n,x}$ is a bijection from $A_{\alpha,\beta,\nu} := \{\pi \in A \,|\, \nu\pi = \alpha\#\beta\}$ onto $A_{\alpha',\beta,\nu} := \{\pi \in A \,|\, \nu\pi = \alpha'\#\beta\}$, because $\pi' = \pi\tau_{n,x}$ is a coplactic neighbour of π whenever $\nu\pi = \alpha\#\beta$. It follows that

$$c_{\alpha,\beta} = \sum_{\nu\in\mathcal{S}^{k.(n-k)}} |A_{\alpha,\beta,\nu}| = \sum_{\nu\in\mathcal{S}^{k.(n-k)}} |A_{\alpha',\beta,\nu}| = c_{\alpha',\beta}\,.$$

Similarly, there is a bijection $\pi \mapsto \pi' := \pi\tau_{n,x+k}$ from $A_{\alpha,\beta,\nu}$ onto $A_{\alpha,\beta',\nu}$ whenever $\beta' = \beta\tau_{n-k,x}$ is a coplactic neighbour of β, which implies that $c_{\alpha,\beta} = c_{\alpha,\beta'}$. The proof is complete. □

8.7 Lemma. *$\mathcal{F}$ is a sub-bialgebra of $\mathcal{Q}$. More precisely, the set of standard Young tableaux of shape F is a union of coplactic classes, for any frame F.*

Proof. Let $n := |F|$ and $\pi \in \mathsf{SYT}^F$. It suffices to show that $\sigma \notin \mathsf{SYT}^F$ implies that σ is not a coplactic neighbour of π, for all $\sigma \in \mathcal{S}_n$.

Let $\sigma \in \mathcal{S}_n$ such that $\sigma \notin \mathsf{SYT}^F$. Assume that $i \in \underline{n-1}$ such that $\sigma = \pi\tau_{n,i}$. (If such an index i does not exist, then σ is not a coplactic neighbour of π anyway.) The choice of π implies that the corresponding mapping $\alpha := \iota_F^{-1}\pi : F \to \underline{n}$ is monotone with respect to $\leq_{\mathbb{Z}\times\mathbb{Z}}$ and $\leq$, while the choice of σ implies that $\tilde{\alpha} := \iota_F^{-1}\sigma = \alpha\tau_{n,i}$ does not have this property. As a consequence, $x := i\alpha^{-1}$ and $y := (i+1)\alpha^{-1}$ are comparable elements of $(F, \leq_{\mathbb{Z}\times\mathbb{Z}})$, and $i \leq i+1$ actually implies $x \leq_{\mathbb{Z}\times\mathbb{Z}} y$. Each cell $z \in F$ with $x \leq_{\mathbb{Z}\times\mathbb{Z}} z \leq_{\mathbb{Z}\times\mathbb{Z}} y$ is equal to x or y, since application of α yields $i \leq z\alpha \leq i+1$, hence $z\alpha \in \{i, i+1\}$. It follows that x and y are horizontal or vertical neighbours in F, that is: $y = (r, c+1)$ or $y = (r+1, c)$, where $r, c \in \mathbb{N}$ such that $x = (r, c)$.

If x and y are horizontal neighbours in F, that is $y = (r, c+1)$, then $x \to y$, hence

$$\langle i, i+1\rangle_\pi = \{\, j \in \underline{n} \,|\, x \to j\alpha^{-1} \to y \,\} = \{i, i+1\}$$

does not contain $i-1$ or $i+2$. So σ is not a coplactic neighbour of π as desired.

Assume now that x and y are vertical neighbours in F, that is $y = (r+1, c)$, then $y \to x$. Let $j \in \langle i, i+1\rangle_\pi \backslash \{i, i+1\}$ and choose $u, v \in \mathbb{N}$ such that $j\alpha^{-1} = (u, v)$. Then $y \to j\alpha^{-1} \to x$ implies $u = r+1$ and $c < v$, or $u = r$ and $v < c$.

In the first case, $x = (r, c) \leq_{\mathbb{Z}\times\mathbb{Z}} (r, v) \leq_{\mathbb{Z}\times\mathbb{Z}} (r+1, v) = j\alpha^{-1}$, hence $(r, v) \in F$, since F is convex. Furthermore, $i \leq (r, v)\alpha \leq j$ and thus $j > i+2$, since $(r, v) \notin \{x, y\}$.

Similarly, in the second case, $j\alpha^{-1} \leq_{\mathbb{Z}\times\mathbb{Z}} (r+1, v) \leq_{\mathbb{Z}\times\mathbb{Z}} y$ and thus $(r+1, v) \in F$. This implies $j \leq (r+1, v)\alpha \leq i+1$. From $(r+1, v) \notin \{x, y\}$ it follows that $j < i-1$.

In other words, in any of the two cases, neither $i-1$ nor $i+2$ is contained in $\langle i, i+1\rangle_\pi$, so that σ is not a coplactic neighbour of π as desired. □

8.8 Corollary. *Let $n \in \mathbb{N}$ and $p \vdash n$, then SYT^p is a coplactic class in $\mathcal{S}_n$ and contained in $\mathcal{G}^p$.*

Proof. $\mathsf{SYT}^p = \mathsf{SYT}^{\mathsf{F}(p)}$ is a union of coplactic classes, by 8.7, and contained in $\mathcal{G}^p$, by 8.2. But each coplactic class A in $\mathcal{G}^p$ has cardinality syt^p, since the restriction to A of the Robinson–Schensted correspondence yields a bijection from A onto $\mathsf{SYT}^p \times \{\beta\}$, where β is the common Q-symbol of the elements of A. □

8.9 Definition. For any linear subspace V of $\mathcal{P}$, the *radical of* V with respect to $(\cdot, \cdot)_\mathcal{P}$ consists of all $\alpha \in V$ such that $(\alpha, \beta)_\mathcal{P} = 0$ for all $\beta \in V$.

8.10 Noncommutative Orthogonality Relations. *Let $n \in \mathbb{N}$, then, for all $p, q \vdash n$ and all coplactic classes A in $\mathcal{G}^p$ and A' in $\mathcal{G}^q$,*

$$(\Sigma A, \Sigma A')_\mathcal{P} = |A' \cap A^{-1}| = \begin{cases} 1 & \text{if } p = q, \\ 0 & \text{otherwise.} \end{cases}$$

In particular, the elements Z^p, $p \vdash n$, span a linear complement of the radical of $\mathcal{Q}_n$ in $\mathcal{Q}_n$, and

$$(\mathrm{Z}^p, \mathrm{Z}^q)_\mathcal{P} = \begin{cases} 1 & \text{if } p = q, \\ 0 & \text{if } p \neq q, \end{cases}$$

for all partitions p, q of n.

Proof. This is a translation of the essentials of the Robinson–Schensted correspondence into linear algebraic properties of $\mathcal{Q}$.

Let $(\Sigma A, \Sigma A')_{\mathcal{P}} \neq 0$ and choose a permutation $\alpha \in A' \cap A^{-1}$. Then $\alpha \in A'$ implies $P(\alpha) \in \mathsf{SYT}^q$ and $\alpha^{-1} \in A$ implies $P(\alpha) = Q(\alpha^{-1}) \in \mathsf{SYT}^p$, hence $p = q$.

Conversely, if $p = q$ is assumed, then there exists a unique permutation $\pi \in \mathcal{S}_n$ such that $Q(\pi)$ is the common Q-symbol of the elements of A' and $P(\pi) = Q(\pi^{-1})$ is the common Q-symbol of the elements of A. Equivalently, $\pi \in A' \cap A^{-1}$. This proves the first claim and implies the orthogonality relations for the elements Z^p, by 8.8.

Furthermore, if $p = q$, then the scalar product $(\Sigma A - \Sigma A', \Sigma B)_{\mathcal{P}}$ is zero for all coplactic classes B in $\mathcal{S}_n$. Thus, on the one hand, the elements $\mathsf{Z}^p - \Sigma A$, where $p \vdash n$ and A is a coplactic class $\neq \mathsf{SYT}^p$ in $\mathcal{G}^p$, constitute a linearly independent subset of the radical of $\mathcal{Q}_n$. On the other hand, $\{ \mathsf{Z}^p \mid p \vdash n \}$ is an orthonormal subset of $\mathcal{Q}_n$. Comparing dimensions, completes the proof. □

In concluding this chapter, an immediate consequence of another result of Schensted [Sch61, Lemma 7] is stated.

8.11 Definition and Remark. Let $p = p_1 \ldots . p_l$ be a partition in $\mathbb{N}^*$. The *conjugate partition* p' of p then has length p_1 and is defined by

$$p'_i = |\{ j \in \underline{l} \mid p_i \geq j \}$$

for all $i \in \underline{p_1}$. In illustrative terms, the partition frame $\mathsf{F}(p')$ corresponding to p' is obtained by transposing the partition frame $\mathsf{F}(p)$ corresponding to p, that is,

$$\mathsf{F}(p') = \{ (y, x) \mid (x, y) \in \mathsf{F}(p) \}.$$

Recall from 5.9 that $\rho_n = n\,(n-1)\,(n-2) \cdots 2\,1$ denotes the order reversing involution in $\mathcal{S}_n$, hence that $\rho_n \pi$ (when viewed as a word) is obtained by reading π backwards, for all $\pi \in \mathcal{S}_n$.

8.12 Theorem. *Let $n \in \mathbb{N}$ and $p \vdash n$. If A is a coplactic class in $\mathcal{G}^p$, then $\rho_n A$ is a coplactic class in $\mathcal{G}^{p'}$.*

A proof of this result may also be found in Appendix C.

Chapter 9

The Main Theorem

In order to complete the noncommutative superstructure, a proof and some immediate consequences are given of Main Theorem 1.3.

A series of primitive elements $\omega_n \in \mathcal{Q}_n$ ($n \in \mathbb{N}$) is then constructed such that $(\omega_n, \omega_n)_{\mathcal{P}} = n$ and $(\omega_n, \mathsf{id}_n)_{\mathcal{P}} = 1$ for all $n \in \mathbb{N}$ (see 9.4 and 9.5). The corresponding algebra map

$$c = c_{\omega} : \mathcal{P} \to \mathcal{C},$$

referred to as Jöllenbeck epimorphism here, will be used for applications in the third part of this book. It is this series upon which the noncommutative theory introduced in [Jöl98] was based.

We shall also address the question in which way the various epimorphisms $c_{\Pi} : \mathcal{Q} \to \mathcal{C}$ may differ as the underlying series of primitive elements Π varies (see 9.8).

9.1 Main Theorem. *Let Π_n be a primitive element in $\mathcal{Q}_n$ such that $(\Pi_n, \Pi_n)_{\mathcal{P}} = n$, for all $n \in \mathbb{N}$. Restricted to $\mathcal{Q}$, the linear map c_{Π} then is a graded and isometric epimorphism of bialgebras. In other words, $c_{\Pi}(\mathcal{Q}_n) = \mathcal{C}\ell_K(\mathcal{S}_n)$ for all $n \in \mathbb{N}_0$, and*

$$(\alpha, \beta)_{\mathcal{P}} = (c_{\Pi}(\alpha), c_{\Pi}(\beta))_{\mathcal{C}}\,,$$
$$c_{\Pi}(\alpha \star \beta) = c_{\Pi}(\alpha) \bullet c_{\Pi}(\beta),$$
$$(c_{\Pi} \otimes c_{\Pi})(\alpha\!\downarrow) = c_{\Pi}(\alpha)\!\downarrow$$

for all $\alpha, \beta \in \mathcal{Q}$.

Proof. Let $n \in \mathbb{N}$ and let C_n be the linear span of the elements Π_p, $p \vdash n$.

From 7.7, it follows that

$$c_\Pi(\Pi_q) = \mathsf{ch}_q$$

for all $q \in \mathbb{N}^*$. The restriction of c_Π to C_n is thus a linear isomorphism onto $\mathcal{C}\ell_K(\mathcal{S}_n)$. Furthermore, $(\Pi_q, \Pi_p)_{\mathcal{P}} = c_\Pi(\Pi_q)(C_p) = \mathsf{ch}_q(C_p) = (\mathsf{ch}_q, \mathsf{ch}_p)_{\mathcal{C}}$, for all partitions q, p of n, hence $c_\Pi|_{C_n}$ is also an isometry.

Denote the radical of $\mathcal{Q}_n$ by R_n and the kernel of $c_\Pi|_{\mathcal{Q}_n}$ by K_n, that is $K_n = \mathcal{Q}_n \cap \ker c_\Pi$. Each element of R_n is, in particular, orthogonal to any element of $C_n \subseteq \mathcal{Q}_n$. The definition of c_Π thus implies $R_n \subseteq K_n$.

The codimension of R_n in $\mathcal{Q}_n$ is equal to the number of partitions of n, by 8.10, hence equal to the codimension of K_n in $\mathcal{Q}_n$, since c_Π maps even C_n onto $\mathcal{C}\ell_K(\mathcal{S}_n)$. It follows that $R_n = K_n$ and thus $\mathcal{Q}_n = C_n + K_n = C_n + R_n$.

Now let $\alpha, \beta \in \mathcal{Q}_n$ and choose elements $\alpha_1, \beta_1 \in C_n$ and $\alpha_2, \beta_2 \in K_n = R_n$ such that $\alpha = \alpha_1 + \alpha_2$ and $\beta = \beta_1 + \beta_2$, then

$$(c_\Pi(\alpha), c_\Pi(\beta))_{\mathcal{C}} = (c_\Pi(\alpha_1), c_\Pi(\beta_1))_{\mathcal{C}} = (\alpha_1, \beta_1)_{\mathcal{P}} = (\alpha, \beta)_{\mathcal{P}} \,.$$

Combined with 7.4, this shows that the restriction of c_Π to the coplactic algebra $\mathcal{Q}$ is an isometric epimorphism of algebras onto $\mathcal{C}$. Since the bilinear form $(\cdot,\cdot)_{\mathcal{C}}$ is regular, an application of 2.14 completes the proof. □

9.2 Remark. In the situation of the Main Theorem, the coefficient $k_n = (\Pi_n, \mathsf{id}_n)_{\mathcal{P}}$ of id_n in Π_n is either 1 or -1.

This is clear for $n = 1$, while for $n > 1$, we use the Main Theorem and 7.5 to obtain inductively

$$\begin{aligned} n! &= n!(\Xi^n, \Xi^n)_{\mathcal{P}} \\ &= n!\Big(c_\Pi(\Xi^n)\,,\, c_\Pi(\Xi^n)\Big)_{\mathcal{C}} \\ &= \sum_{p \vdash n} |C_p| \left(c_\Pi(\Xi^n)(C_p)\right)^2 \\ &= \sum_{p \neq n} |C_p| + |C_n|\, k_n^2 \\ &= n! + |C_n|\,(k_n^2 - 1), \end{aligned}$$

hence also $k_n = 1$ or $k_n = -1$.

The following additional consequence of 8.10 is worth mentioning.

9.3 Corollary. $c_\Pi(\Sigma A) = c_\Pi(Z^p)$, *for any partition p in $\mathbb{N}^*$ and any coplactic class A contained in the Greene cell $\mathcal{G}^p$.*

Proof. The difference $Z^p - \Sigma A$ is orthogonal to each element of $\mathcal{Q}$, by 8.10, hence, in particular, to each element Π_p, where p is a partition in $\mathbb{N}^*$. □

One possible choice for Π_n in the Main Theorem is the following.

9.4 Definition and Remarks. For all $n \in \mathbb{N}$, set

$$\omega_n := \sum_{k=0}^{n-1} (-1)^k Z^{(n-k).1^k},$$

where $(n-k).1^k := (n-k).\underbrace{1.1.\ldots.1}_{k} \vdash n$ for all $k \in \underline{n-1}_| \cup \{0\}$. For example, viewing permutations as words, $\omega_3 = Z^3 - Z^{2.1} + Z^{1.1.1} = 1\,2\,3 - 2\,1\,3 - 3\,1\,2 + 3\,2\,1$. By definition, $\omega_n \in \mathcal{F}$. Considering the second basis of Solomon's algebra mentioned in the introduction, $Z^{(n-k).1^k} = \Delta^{\underline{k}_|}$ is the sum of all permutations $\pi \in \mathcal{S}_n$ with descent set $\underline{k}_|$, for all $k \in \underline{n-1}_| \cup \{0\}$ (see also 11.6 and 11.7), hence ω_n is actually contained in $\mathcal{D}_n$.

Furthermore, the identity $Z^{1^k} \star Z^m = Z^{m.1^k} + Z^{(m+1).1^{k-1}}$ mentioned in 6.10, for all $k, m \in \mathbb{N}$, allows the definition of ω_n to be rewritten as

$$\omega_n = \sum_{k=0}^{n} (-1)^k (n-k) Z^{1^k} \star Z^{n-k}.$$

9.5 Theorem. *For each $n \in \mathbb{N}$, ω_n is a primitive element of $K\mathcal{S}_n$ with $(\omega_n, \omega_n)_{\mathcal{P}} = n$ and $(\omega_n, \mathrm{id}_n)_{\mathcal{P}} = 1$.*

Proof. The coefficient of id_n in ω_n is 1, by definition.

Recall from 6.10 that $A_m := \sum_{k=0}^{m} (-1)^k Z^{1^k} \star Z^{m-k} = 0$ for all $m > 0$, while $A_0 = \emptyset$. Apply 5.11 and the restriction rules for $Z^n = \mathrm{id}_n$ and $Z^{1^n} = \rho_n$ stated in 5.9 to obtain

$$\begin{aligned}
\omega_n\!\downarrow &= \sum_{k=0}^{n} (-1)^k (n-k)\, (Z^{1^k}\!\downarrow) \star_{\otimes} (Z^{n-k}\!\downarrow) \\
&= \sum_{k=0}^{n} (-1)^k (n-k) (\sum_{i=0}^{k} Z^{1^i} \otimes Z^{1^{k-i}}) \star_{\otimes} (\sum_{j=0}^{n-k} Z^j \otimes Z^{n-k-j}) \\
&= \sum_{i=0}^{n} \sum_{j=0}^{n-i} \sum_{k=i}^{n-j} (-1)^k (n-k)\, (Z^{1^i} \star Z^j) \otimes (Z^{1^{k-i}} \star Z^{n-k-j}) \\
&= \sum_{m=0}^{n} \sum_{i=0}^{m} \sum_{k=i}^{n-m+i} (-1)^k (n-k)\, (Z^{1^i} \star Z^{m-i}) \otimes (Z^{1^{k-i}} \star Z^{n-k-m+i})
\end{aligned}$$

$$= \sum_{m=0}^{n}\sum_{i=0}^{m}\sum_{k=0}^{n-m}(-1)^{k+i}(n-k-i)\,(\mathrm{Z}^{1^i}\star\mathrm{Z}^{m-i})\otimes(\mathrm{Z}^{1^k}\star\mathrm{Z}^{n-k-m})$$

$$= \sum_{m=0}^{n}\Big((\sum_{i=0}^{m}(-1)^i(m-i)\,\mathrm{Z}^{1^i}\star\mathrm{Z}^{m-i})\otimes(\sum_{k=0}^{n-m}(-1)^k\,\mathrm{Z}^{1^k}\star\mathrm{Z}^{n-m-k})$$

$$+(\sum_{i=0}^{m}(-1)^i\,\mathrm{Z}^{1^i}\star\mathrm{Z}^{m-i})\otimes(\sum_{k=0}^{n-m}(-1)^k(n-m-k)\,\mathrm{Z}^{1^k}\star\mathrm{Z}^{n-m-k})\Big)$$

$$= \sum_{m=0}^{n}(\omega_m\otimes A_{n-m}+A_m\otimes\omega_{n-m})$$

$$= \omega_n\otimes\emptyset+\emptyset\otimes\omega_n\,.$$

This proves the first claim and implies that

$$(\omega_n,\omega_n)_{\mathcal{P}} = \sum_{k=0}^{n}(-1)^k(n-k)(\omega_n,\mathrm{Z}^{1^k}\star\mathrm{Z}^{n-k})_{\mathcal{P}} = n(\omega_n,\mathrm{Z}^n)_{\mathcal{P}} = n,$$

as asserted. □

9.6 Definition and Remarks. Set $\omega_q = \omega_{q_1}\star\cdots\star\omega_{q_k}$ for all $q = q_1.\dots.q_k \in \mathbb{N}^*$. The Main Theorem can be applied to the *Jöllenbeck epimorphism* $c = c_{\omega} : \mathcal{P}\to\mathcal{C}$, defined by

$$c(\varphi)(C_p) = (\varphi,\omega_p)_{\mathcal{P}}$$

for all $n\in\mathbb{N}$, $\varphi\in K\mathcal{S}_n$ and $p\vdash n$, which extends Solomon's epimorphism $c_n : \mathcal{D}_n\to \mathcal{C}\ell_K(\mathcal{S}_n)$ for each $n\in\mathbb{N}$, by 7.6. Furthermore,

$$c(\omega_q) = \mathsf{ch}_q$$

for all $q\in\mathbb{N}^*$, by 7.7.

Recall from 2.9 that δ denotes the unique coproduct on $K\mathbb{N}^*$ such that $(K\mathbb{N}^*, ., \delta)$ is a bialgebra and $n\delta = n\otimes\varnothing+\varnothing\otimes n$ for all $n\in\mathbb{N}$.

9.7 Corollary. *$\{\,\omega_q\,|\,q\in\mathbb{N}^*\,\}$ is a linear basis of $\mathcal{D}$. In particular, the map $q\mapsto\omega_q$ $(q\in\mathbb{N}^*)$ extends linearly to an isomorphism of bialgebras*

$$(K\mathbb{N}^*, ., \delta)\longrightarrow(\mathcal{D},\star,\downarrow).$$

Proof. Let $n\in\mathbb{N}$ and let $\omega_n = \sum_{q\models n} a_q\Xi^q$ with coefficients $a_q\in K$, then

$$n = (\omega_n,\omega_n)_{\mathcal{P}} = \sum_{q\models n} a_q(\Xi^q,\omega_n)_{\mathcal{P}} = a_n\,,$$

by 9.5 and 5.14. The basis property of $\{\,\omega_q \mid q \in \mathbb{N}^*\,\}$ thus follows from the basis property of $\{\,\Xi^q \mid q \in \mathbb{N}^*\,\}$, by triangularity. The second claim now also follows from 9.5. □

9.8 Proposition. *Suppose that* Π_n *is a primitive element in* $\mathcal{Q}_n$ *such that* $(\Pi_n, \Pi_n)_{\mathcal{P}} = n$ *and* $(\Pi_n, \mathrm{id}_n)_{\mathcal{P}} = 1$*, for each* $n \in \mathbb{N}$*. Then* c_Π *and* c *coincide on* $\mathcal{Q}$.

Proof. Let $n \in \mathbb{N}$. The kernel of both c and c_Π in $\mathcal{Q}_n$ is equal to R_n, the intersection of $\mathcal{Q}_n$ with the radical of $\mathcal{Q}$. Furthermore, the elements Π_p, $p \vdash n$, span a linear complement of R_n in $\mathcal{Q}_n$. This was shown in the proof of 9.1.

Primitivity of Π_n implies that $c(\Pi_n)(C_p) = 0$ for all partitions p of n of length > 1, while

$$\begin{aligned} c(\Pi_n)(C_n) &= (\omega_n, \Pi_n)_{\mathcal{P}} \\ &= \sum_{k=0}^{n} (-1)^k (n-k)(\mathrm{Z}^{1^k} \star \mathrm{Z}^{n-k}, \Pi_n)_{\mathcal{P}} \\ &= n(\mathrm{Z}^n, \Pi_n)_{\mathcal{P}} \\ &= n. \end{aligned}$$

This shows $c(\Pi_n) = \mathsf{ch}_n$, hence $c(\Pi_q) = \mathsf{ch}_q = c_\Pi(\Pi_q)$ for all $q \in \mathbb{N}^*$. The claim follows. □

This gives at hand a huge variety of descriptions of the map c. A linear basis of the space of all primitive elements in $\mathcal{Q}_n$, however, is not known. The space of primitive elements in $\mathcal{D}_n$ is $\omega_n \mathcal{D}_n$ (see, for instance, [Sch, Main Theorem 7 and its proof]).

To conclude this chapter, recall that the Solomon descent algebra $\mathcal{D}_n$ is a subalgebra of the group algebra $K\mathcal{S}_n$ and that c yields a homomorphism of algebras onto the ring of class functions $\mathcal{C}\ell_K(\mathcal{S}_n)$ when restricted to $\mathcal{D}_n$, by Theorem 1.1.

Little is known about products in the group algebra $K\mathcal{S}_n$ of the elements of the larger subspace $\mathcal{Q}_n$ — except for the fact that they are not contained in $\mathcal{Q}_n$ in general, so that $\mathcal{Q}_n$ is *not* a subalgebra of $K\mathcal{S}_n$. In fact, not even the product of $\Xi^{1.3} \in \mathcal{D}_4$ with $\mathrm{Z}^{2.2} \in \mathcal{Q}_4$ lies in $\mathcal{Q}_4$, as is readily verified using 8.5. There is, however, the following generalisation of Solomon's homomorphism rule.

9.9 Theorem. *$c(\alpha\beta) = c(\alpha)c(\beta) = c(\beta\alpha)$, for all $n \in \mathbb{N}$, $\alpha \in \mathcal{D}_n$ and $\beta \in \mathcal{Q}_n$.*

Proof. If both α and β are contained in $\mathcal{D}_n$, then this is a special case of Solomon's theorem.

Assume that β is an arbitrary element of $\mathcal{Q}_n$. The associativity law $(\alpha\beta, \gamma)_{\mathcal{P}} = (\alpha, \beta\gamma)_{\mathcal{P}}$ holds for permutations α, β and γ, simply by definition of $(\cdot,\cdot)_{\mathcal{P}}$, and then extends to arbitrary elements α, β, γ of $\mathcal{P}$, by linearity. As a consequence, for all partitions p of n,

$$\begin{aligned} c(\alpha\beta)(C_p) &= (\alpha\beta, \omega_p)_{\mathcal{P}} \\ &= \Big(\beta\,,\, \omega_p\alpha\Big)_{\mathcal{P}} \\ &= \Big(c(\beta)\,,\, c(\omega_p)c(\alpha)\Big)_{\mathcal{C}} \\ &= \Big(c(\alpha)c(\beta)\,,\, \mathsf{ch}_p\Big)_{\mathcal{C}} \\ &= \Big(c(\alpha)c(\beta)\Big)(C_p), \end{aligned}$$

by the Main Theorem, since $\omega_p\alpha \in \mathcal{D}_n \subseteq \mathcal{Q}_n$. The identity $c(\alpha\beta) = c(\alpha)c(\beta)$ follows. An analogous argument shows $c(\beta\alpha) = c(\alpha)c(\beta)$. □

PART III

Classical Character Theory of the Symmetric Group

Chapter 10

Irreducible Characters

Let $n \in \mathbb{N}$. The noncommutative orthogonality relations 8.10 suggest that the elements Z^p, $p \vdash n$, of $\mathcal{Q}_n$ be considered as noncommutative irreducible characters of $\mathcal{S}_n$. Indeed, the isometry property of the epimorphism c stated in the Main Theorem implies that the class functions

$$\zeta^p := c(Z^p)$$

form an orthonormal subset of $\mathcal{C}\ell_K(\mathcal{S}_n)$ indexed by partitions p of n. However, the problem remains to show that these class functions are actually characters. One way to solve this problem is to establish a link to (noncommutative) Young characters, as follows.

The lexicographic order on $\mathbb{N}^*$ is denoted by $\leq_{\text{lex}}$.

10.1 Lemma. *Let* $n \in \mathbb{N}$ *and* $p, q \vdash n$, *then* $(\Xi^q, Z^p)_{\mathcal{P}} = 0$ *unless* $q \leq_{lex} p$, *and* $(\Xi^p, Z^p)_{\mathcal{P}} = 1$.

Proof. If $q = n$, then this is part of 8.10, since $\Xi^n = \mathrm{id}_n = Z^n$. Assume that $q = q_1. \dots .q_l$ is of length $l > 1$, and set $\tilde{q} := q_2. \dots .q_l$. The self-duality 5.14 and the restriction rule 6.13 then yield that

$$\begin{aligned}
(\Xi^q, Z^p)_{\mathcal{P}} &= (\Xi^{q_1} \star \Xi^{\tilde{q}}, Z^p)_{\mathcal{P}} \\
&= (\Xi^{q_1} \otimes \Xi^{\tilde{q}}, Z^p{\downarrow})_{\mathcal{P}\otimes\mathcal{P}} \\
&= \sum_{r \subseteq p} (\Xi^{q_1}, Z^r)_{\mathcal{P}} (\Xi^{\tilde{q}}, Z^{p \setminus r})_{\mathcal{P}} \\
&= \sum_{r \subseteq p} (Z^{q_1}, Z^r)_{\mathcal{P}} (\Xi^{\tilde{q}}, Z^{p \setminus r})_{\mathcal{P}} .
\end{aligned}$$

This is zero unless the partition $r = q_1$ of length one is contained in p, that is, unless $q_1 \leq p_1$. Therefore $(\Xi^q, \mathrm{Z}^p)_{\mathcal{P}} \neq 0$ implies that either $q_1 < p_1$, hence $q \leq_{\mathrm{lex}} p$, or $q_1 = p_1$ and

$$(\Xi^q, \mathrm{Z}^p)_{\mathcal{P}} = (\Xi^{\tilde{q}}, \mathrm{Z}^{p \setminus p_1})_{\mathcal{P}} = (\Xi^{\tilde{q}}, \mathrm{Z}^{\tilde{p}})_{\mathcal{P}},$$

where $\tilde{p} = p_2. \dots .p_k$. In the second case, the proof may be completed by a simple induction on l. □

The following important result can now be stated and proven.

10.2 Theorem. *Let $n \in \mathbb{N}_0$, then $\{\zeta^p \,|\, p \vdash n\}$ is the set of irreducible $\mathcal{S}_n$-characters, and $\zeta^p \neq \zeta^q$ whenever $p \neq q$. Furthermore,*

$$\deg \zeta^p = \mathsf{syt}^p \quad \textit{and} \quad (\zeta^p, \zeta^q)_{\mathcal{C}} = \begin{cases} 1 & \textit{if } p = q, \\ 0 & \textit{if } p \neq q, \end{cases}$$

for all $p, q \vdash n$.

Proof. As already mentioned, using the Main Theorem, the noncommutative orthogonality relations 8.10 become

$$(\zeta^p, \zeta^q)_{\mathcal{C}} = (\mathrm{Z}^p, \mathrm{Z}^q)_{\mathcal{P}} = \begin{cases} 1 & \text{if } p = q, \\ 0 & \text{if } p \neq q, \end{cases}$$

for all $p, q \vdash n$. In particular, $\zeta^p \neq \zeta^q$ whenever $p \neq q$. More precisely, comparing dimensions, the elements ζ^p, $p \vdash n$, constitute an orthonormal basis of $\mathcal{C}\ell_K(\mathcal{S}_n)$.

We employ the Young characters in order to show that ζ^p is actually a character for all $p \vdash n$. Let $k_{qp} \in K$ be so chosen that

$$\xi^q = \sum_{p \vdash n} k_{qp} \zeta^p,$$

for all $q, p \vdash n$, that is, $k_{qp} = (\xi^q, \zeta^p)_{\mathcal{C}} = (\Xi^q, \mathrm{Z}^p)_{\mathcal{P}}$ for all $q, p \vdash n$, by orthonormality and 7.6. Assuming lexicographically increasing order of row and column indices, it follows from 10.1 that $K_n := (k_{qp})_{q,p \vdash n}$ is an upper uni-triangular matrix over $\mathbb{Z}$. The determinant of K_n is thus 1, and K_n is invertible over $\mathbb{Z}$. Since ξ^q is a character and thus a linear combination of irreducible $\mathcal{S}_n$-characters with coefficients in $\mathbb{N}_0$, for all $q \vdash n$, it follows that ζ^p is an integer linear combination of irreducible characters, for all $p \vdash n$. Now $(\zeta^p, \zeta^p)_{\mathcal{P}} = 1$ implies that, for all p, either ζ^p or $-\zeta^p$ is irreducible.

But

$$\zeta^p(\mathrm{id}_n) = (Z^p, \omega_{1^n})_{\mathcal{P}} = (Z^p, \Xi^{1^n})_{\mathcal{P}} = \Big(Z^p, \sum_{\pi\in S_n} \pi\Big)_{\mathcal{P}} = \mathrm{syt}^p$$

and hence, by positivity of degree, it follows that ζ^p is an irreducible character, not $-\zeta^p$. □

The preceding theorem implies:

10.3 Corollary. *$\mathbb{Q}$ is a splitting field of S_n, for all $n \in \mathbb{N}_0$.*

Note that, by 9.3, the image under c of the sum of *any* coplactic class A in S_n is an irreducible S_n-character, namely the character ζ^p, where $p \vdash n$ such that A is contained in the Greene cell $\mathcal{G}^p$. More generally, the following result provides a perfect tool for applications.

10.4 Corollary. *Let $n \in \mathbb{N}_0$, and let $Y \subseteq S_n$ such that $\Sigma Y \in \mathcal{Q}$ or, equivalently, Y is a union of coplactic classes, then $c(\Sigma Y)$ is a character of S_n of degree $|Y|$. Furthermore, for all $p \vdash n$, the multiplicity of ζ^p in $c(\Sigma Y)$ is equal to*

$$(c(\Sigma Y), \zeta^p)_{\mathcal{C}} = \mathrm{syt}^p(Y) := |Y^{-1} \cap \mathrm{SYT}^p|,$$

which is the number of standard Young tableaux π of shape p such that $\pi^{-1} \in Y$.

Proof. Applying 10.2 and the Main Theorem,

$$c(\Sigma Y) = \sum_{p\vdash n}(\zeta^p, c(\Sigma Y))_{\mathcal{C}}\, \zeta^p = \sum_{p\vdash n}(Z^p, \Sigma Y)_{\mathcal{P}}\, \zeta^p = \sum_{p\vdash n} |\mathrm{SYT}^p \cap Y^{-1}|\, \zeta^p,$$

by the definition of the scalar product on $\mathcal{P}$. Hence $c(\Sigma Y)$ is a character. The degree of $c(\Sigma Y)$ is

$$\deg c(\Sigma Y) = c(\Sigma Y)(\mathrm{id}_n) = (\Sigma Y, \omega_{1^n})_{\mathcal{P}} = \Big(\Sigma Y, \sum_{\pi\in S_n} \pi\Big)_{\mathcal{P}} = |Y|$$

as asserted. □

10.5 Remark. For any submodule M of the regular S_n-module there is a suitable union Y of coplactic classes in S_n such that ΣY is a noncommutative character corresponding to χ_M. Indeed, for all $p \vdash n$,

$$a_p := (\chi_M, \zeta^p)_{\mathcal{C}} \leq (\chi_{KS_n}, \zeta^p)_{\mathcal{C}} = \mathrm{syt}^p,$$

thus by 8.3, a_p pairwise disjoint coplactic classes A_i^p, $i \in \underline{a_p}\rfloor$, in $\mathcal{G}^p$ may be chosen. Then the sum of the union

$$Y := \bigcup_{p \vdash n} \bigcup_{i=1}^{a_p} A_i^p$$

is a noncommutative character corresponding to χ_M, since $c(\Sigma A_i^p) = \zeta^p$ for all p and i.

However, in view of applications, the problem is to find a noncommutative character of this type *without* complete knowledge of the decomposition numbers of M. As a consequence of 8.7, such a noncommutative character is given, almost by definition, for the huge class of so-called skew characters of $\mathcal{S}_n$ (see the examples below).

10.6 Examples. Let F be a frame of order n and let p, q be partitions according to 6.6 such that $F \simeq \mathsf{F}(p\backslash q)$, then the set $Y = \mathsf{SYT}^F = \mathsf{SYT}^{p\backslash q}$ is a union of coplactic classes, by 8.7. Hence, by 10.4,

$$\zeta^F = \zeta^{p\backslash q} := c(\mathrm{Z}^F) = c(\Sigma Y),$$

is a character of $\mathcal{S}_n$ of degree $\mathsf{syt}^F := |\mathsf{SYT}^F|$, called a *skew character.* Consider two extreme cases:

(i) Let $Y = \mathcal{S}_n = \mathsf{SYT}^{\mathsf{HS}(1^n)}$, then

$$\Sigma Y = \Xi^{1^n} = \sum_{p \vdash n} \Sigma \mathcal{G}^p = \sum_{p \vdash n} \quad \sum_{A \text{ coplactic class in } \mathcal{G}^p} \Sigma A$$

is a noncommutative character of $\mathcal{S}_n$, corresponding to the regular character $\chi_{K\mathcal{S}_n}$. In particular,

$$\chi_{K\mathcal{S}_n} = \sum_{p \vdash n} \mathsf{syt}^p \, \zeta^p.$$

If $p \vdash n$ and M_p denotes an irreducible $\mathcal{S}_n$-module with character ζ^p, then $\Sigma \mathcal{G}^p$ is a noncommutative character corresponding to the homogeneous component H_p of $K\mathcal{S}_n$ of type M_p. Furthermore, we have $\Sigma \mathcal{G}^p = \sum_A \Sigma A$, where the sum is taken over all coplactic classes in $\mathcal{G}^p$. This reflects the decomposition of H_p into syt^p irreducible submodules isomorphic to M_p.

(ii) Let $n \in \mathbb{N}_0$. The case where $Y = \{\mathsf{id}_n\}$ was already analysed in 7.5. Let $Y := \{\rho_n\}$, then we have $\Sigma Y = \rho_n = \mathrm{Z}^{1^n}$. Hence

$$\mathsf{sgn}_n := \zeta^{1^n} = c(\rho_n)$$

is a linear character of $\mathcal{S}_n$, that is, a character of degree 1. This character is the *sign character* of $\mathcal{S}_n$. As ρ_n is an involution, the value $\mathsf{sgn}_n(C_p) = (\rho_n, \omega_p)_\mathcal{P}$ of sgn_n is the coefficient of ρ_n in ω_p, for all $p \vdash n$. If $p = n$, this coefficient is $(-1)^{n-1}$, by definition of ω_n. As for id_n, using the coproduct rule for ρ_n given in 5.9, it follows by induction on the length of q that

$$\mathsf{sgn}_n(C_q) = (-1)^{n-\ell(q)},$$

for all $q \vdash n$.

In addition to the irreducible characters ζ^p (see Chapters 11 and 13), important examples of skew characters are given by the family of *Young characters* ξ^q (see Chapter 12) and the family of *descent characters* δ^q (see Chapter 14).

Note that the noncommutative sign character ρ_n is the unique element in $\mathcal{S}_n$ with maximal descent set $\underline{n-1}$, hence actually contained in $\mathcal{D}_n$. This allows one to apply 9.9 and to translate Schensted's theorem 8.12 as follows.

10.7 Corollary. $\mathsf{sgn}_n\zeta^p = \zeta^{p'}$, *for all* $n \in \mathbb{N}$ *and all partitions* p *of* n.

Proof. It follows from 8.8 and 8.12 that $\rho_n\mathsf{SYT}^p$ is a coplactic class in $\mathcal{G}^{p'}$. Applying 9.9 and 9.3, we get $\mathsf{sgn}_n\zeta^p = c(\rho_n)c(\mathrm{Z}^p) = c(\rho_n\mathrm{Z}^p) = \zeta^{p'}$. □

Chapter 11

The Murnaghan–Nakayama Rule

This result, stated and proved for the more general setting of arbitrary skew characters, provides a recursive formula for the values of the irreducible characters of any symmetric group $\mathcal{S}_n$. Throughout this chapter, $n \in \mathbb{N}_0$ is fixed.

Let F be a frame of order n and $q \vdash n$, then, by definition, the value of the skew character ζ^F of $\mathcal{S}_n$ on each element of the conjugacy class C_q is given by

$$\zeta^F(C_q) = c(\mathrm{Z}^F)(C_q) = (\mathrm{Z}^F, \omega_q)_{\mathcal{P}} .$$

The more explicit recursive formula for these values mentioned above is obtained by applying the restriction rule 5.10 and the reciprocity law 5.14. For the sake of an example, this chapter begins with a short investigation in the behaviour of an irreducible character of $\mathcal{S}_n$ under restriction to $\mathcal{S}_{n-1}$, or, more precisely, to the stabiliser $\mathcal{S}_{(n-1).1}$ of n in $\mathcal{S}_n$.

11.1 Branching Rule. *Let $p \vdash n$, then*

$$\zeta^p|_{\mathcal{S}_{n-1}} = \sum_q \zeta^q,$$

where the sum ranges over all partitions q of $n-1$ such that $q \subseteq p$.

Proof. As mentioned in 3.9, the conjugacy classes of $\mathcal{S}_{(n-1).1}$ are $C_r \# C_1$, $r \vdash n-1$, and $C_r \# C_1 \subseteq C_{r.1}$. It follows that

$$\begin{aligned}\zeta^p|_{\mathcal{S}_{(n-1).1}}(C_r \# C_1) &= \zeta^p(C_{r.1}) \\ &= (\mathrm{Z}^p, \omega_r \star \omega_1)_{\mathcal{P}} \\ &= (\mathrm{Z}^p\!\downarrow, \omega_r \otimes \omega_1)_{\mathcal{P}\otimes\mathcal{P}}\end{aligned}$$

$$= \sum_{q \subseteq p} (\mathsf{Z}^q, \omega_r)_{\mathcal{P}} (\mathsf{Z}^{p \setminus q}, \omega_1)_{\mathcal{P}}$$

$$= \sum_{\substack{q \vdash n-1 \\ q \subseteq p}} (\mathsf{Z}^q, \omega_r)_{\mathcal{P}}$$

$$= \sum_{\substack{q \vdash n-1 \\ q \subseteq p}} \zeta^q(C_r),$$

for all $r \vdash n-1$, by 5.14 and 6.13. □

We turn to the general case.

11.2 Definition. Let $r = r_1. \dots .r_l \in \mathbb{N}^*$ and let $s_i := r_1 + \dots + r_i$, for all $i \in \underline{l}$. Put $p := (s_l - (l-1)).(s_{l-1} - (l-2)). \dots .(s_2 - 1).s_1 \in \mathbb{N}^*$ and $q := (s_{l-1} - (l-1)). \dots .(s_2 - 2).(s_1 - 1) \in \mathbb{N}^*$, omitting the zeros. Then p and q are partitions. The partition frame $\mathsf{F}(p)$ has width p_1 and height l and contains the frame $H := \mathsf{F}(p_1.1^{l-1})$ which has "hook shape":

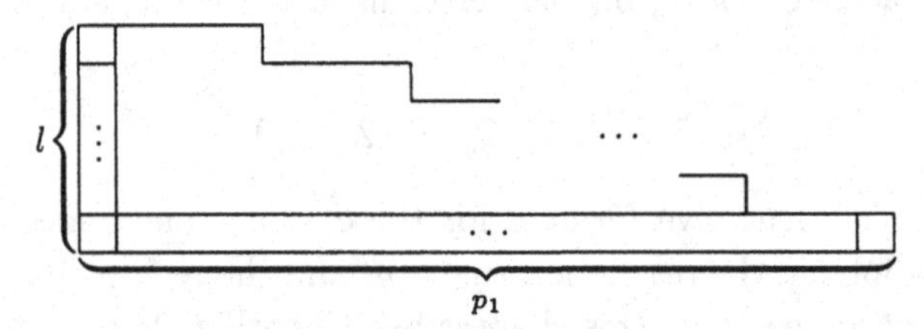

Traditionally, the corner cell $(1,1)$ of H is called its *head*, the vertical bar in H, excluding the head, is called its *leg* and the horizontal bar in H, excluding the head, is called its *arm**.

Each cell in H corresponds to a unique cell in the rim

$$\mathsf{RH}(r) := \mathsf{F}(p) \setminus \mathsf{F}(q)$$

of $\mathsf{F}(p)$ which may be visualised by

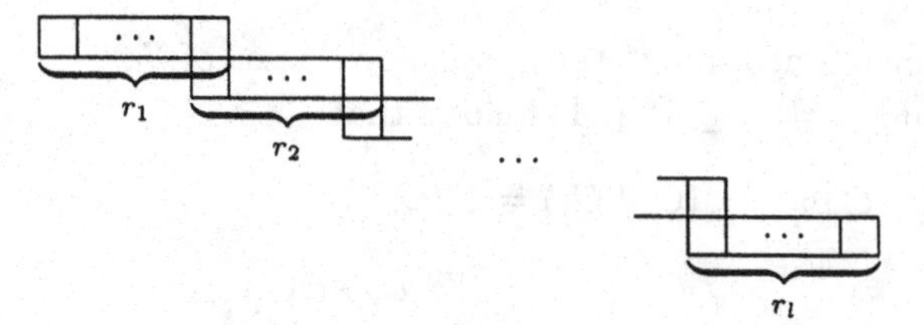

*With regard to our visualisation one might think of a ballet dancer here.

Any frame F such that $F \simeq \mathsf{RH}(r)$ is called a *rim hook* of type r and

$$\mathsf{leg}(F) := l - 1 = \ell(r) - 1$$

is called its *leg length.*

11.3 Theorem. *Let F be a frame and $q \in \mathbb{N}^*$, then*

$$\zeta^F(C_{q.n}) = \sum_I (-1)^{\mathsf{leg}(F\setminus I)}\, \zeta^I(C_q),$$

where the sum is taken over all ideals I of F such that $F\setminus I$ is a rim hook of order n. In particular,

$$\zeta^F(C_n) = \begin{cases} (-1)^{\mathsf{leg}(F)} & \textit{if } F \textit{ is a rim hook of order } n, \\ 0 & \textit{otherwise.} \end{cases}$$

11.4 Example. In order to calculate the character value $\zeta^{4.4.3\setminus 2}(C_{3.2.2.2})$, we consider the following illustration of the preceding recursive formula:

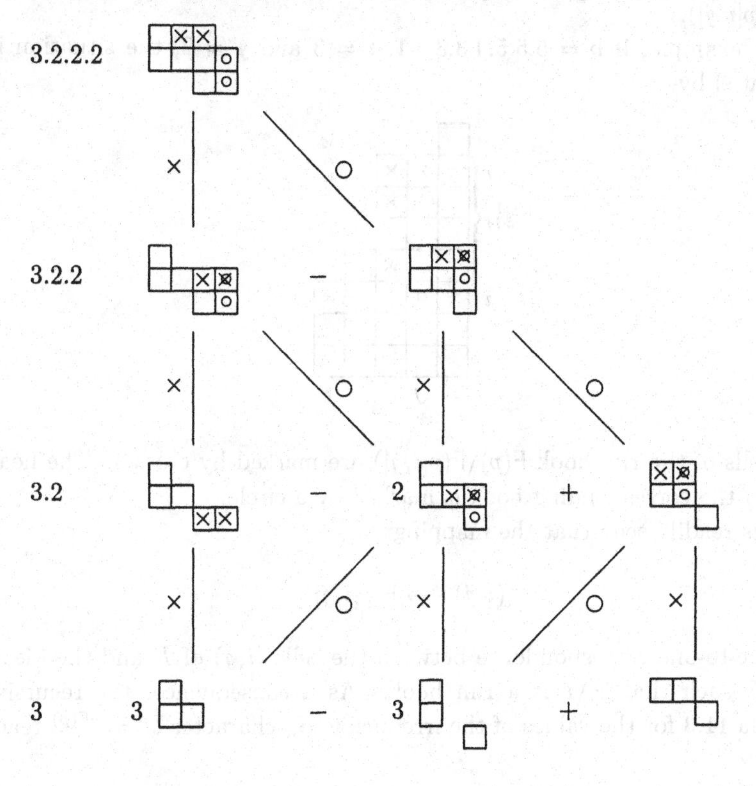

At each step, the rim hooks $\tilde{F}$ with even (respectively, odd) leg lengths are marked by crosses (respectively, circles). Accordingly,

$$\begin{aligned}\zeta^{4.4.3\backslash 2}(C_{3.2.2.2}) &= \zeta^{4.4.1\backslash 2}(C_{3.2.2}) - \zeta^{3.3.3\backslash 2}(C_{3.2.2}) \\ &= \zeta^{4.2.1\backslash 2}(C_{3.2}) - 2\zeta^{3.3.1\backslash 2}(C_{3.2}) + \zeta^{3.2.2\backslash 2}(C_{3.2}) \\ &= 3\zeta^{2.1}(C_3) - 3\zeta^{3.1.1\backslash 2}(C_3) + \zeta^{3.2\backslash 2}(C_3) \\ &= -3.\end{aligned}$$

Before proving the theorem, we consider the partition frame $F = \mathsf{F}(p)$, where p is a partition of k, say. Let I be an ideal of F such that $F\backslash I$ is a rim hook. Choose $i, j, r, s \in \mathbb{N}$ such that (i, s) is maximal and (r, j) is minimal in $F\backslash I$ with respect to $\rightarrow$. Then $(i, j) \in F$,

$$h^p_{i,j} := |F\backslash I| = r - i + s - j + 1 \quad \text{and} \quad \mathsf{leg}^p_{i,j} := \mathsf{leg}(F\backslash I) = r - i.$$

Furthermore, by 6.11, there exists a partition $p[i, j] \vdash k - h^p_{i,j}$ such that $I = \mathsf{F}(p[i, j])$.

For example, if $p = 5.5.5.4.3.3.3.1$, $i = 3$ and $j = 2$, the situation is illustrated by

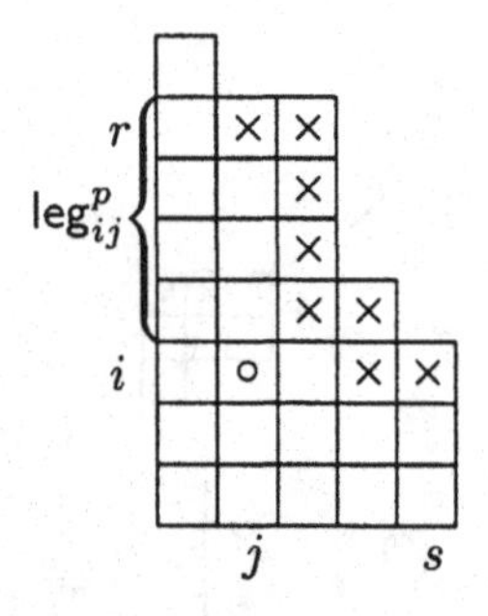

The cells of the rim hook $\mathsf{F}(p)\backslash\mathsf{F}(p[i, j])$ are marked by crosses. The head (i, j) of the corresponding hook is marked by a circle.

It is readily seen that the mapping

$$(i, j) \longmapsto \mathsf{F}(p[i, j])$$

is a one-to-one correspondence between the cells (i, j) of F and the ideals I of F such that $F\backslash I$ is a rim hook. As a consequence, the recursive formula 11.3 for the values of the irreducible $\mathcal{S}_n$-character $\zeta^p = \zeta^{\mathsf{F}(p)}$ reads as follows.

11.5 Murnaghan–Nakayama Rule. *For all $m \in \mathbb{N}$ and all $p \vdash m+n$, $q \models m$, we have*

$$\zeta^p(C_{q.n}) = \sum_{(i,j)} (-1)^{\mathsf{leg}^p_{i,j}} \zeta^{p[i,j]}(C_q),$$

where the sum is taken over all $(i,j) \in \mathsf{F}(p)$ such that $h^p_{i,j} = n$.

The proof of 11.3 requires a short study of the element ω_n and a characterisation of rim hooks.

11.6 Definition and Remark. Let $\pi \in S_n$ and recall that the descent set of π is $\mathsf{Des}(\pi) = \{\, i \in \underline{n-1} \mid i\pi > (i+1)\pi \,\}$. The *number of descents* of π is denoted by $\mathsf{des}(\pi) := |\mathsf{Des}(\pi)|$.

For all $D \subseteq \underline{n-1}$, the *descent class* $\mathcal{S}_n(D)$ corresponding to D consists of all permutations π in S_n such that $\mathsf{Des}(\pi) = D$. Note that if $r = r_1.\dots.r_l \models n$ and $D = \{r_1, r_1 + r_2, \dots, r_1 + \cdots + r_{l-1}\}$, then the descent class $\mathcal{S}_n(D)$ coincides with the set $\mathsf{SYT}^{\mathsf{RH}(r)}$ of standard Young tableaux of shape $\mathsf{RH}(r)$. The elements of

$$\mathcal{V}_n := \bigcup_{0 \leq k < n} \mathcal{S}_n(\underline{k})$$

are the *valley permutations* in S_n.

The valley permutations are associated with an amazing combinatorial world (see, for instance, [BL93]). We restrict ourselves to some immediate results needed here.

11.7 Proposition. *Let $\pi \in \mathcal{V}_n$ and put $k := \mathsf{des}(\pi)$, then:*

(i) $1\pi > 2\pi > \cdots k\pi > (k+1)\pi = 1 < (k+2)\pi < \cdots < n\pi$;
(ii) $\underline{j}\pi^{-1}$ *is an interval in $\underline{n}$, for all $j \in \underline{n}$;*
(iii) $\mathsf{Des}(\pi^{-1}) = \mathsf{Des}(\pi)\pi - 1$.

In particular, the set $\mathcal{V}_n^{-1}$ of inverse valley permutations is a transversal of the descent classes in S_n.

Proof. As π is a valley permutation with k descents, $\mathsf{Des}(\pi) = \underline{k}$, or, equivalently, (i).

For any $j \in \underline{n}$, choose $i \in \underline{k+1}$ minimal such that $i\pi \leq j$, and $l \in \underline{n} \setminus \underline{k}$ maximal such that $l\pi \leq j$, then (i) implies $\underline{j}\pi^{-1}$ is contained in the interval $\langle i, l \rangle$ with margins i and l, proving (ii).

Consider $i \in \underline{n-1}$ and view π as a word. Then $i+1$ stands to the left of $(k+1)\pi = 1$ in π if and only if $i\pi^{-1} > (i+1)\pi^{-1}$. Equivalently, $i+1 \in \underline{k}\pi = \mathsf{Des}(\pi)\pi$ if and only if $i \in \mathsf{Des}(\pi^{-1})$. This proves (iii).

In particular, for any $\sigma \in \mathcal{V}_n$ such that $\mathsf{Des}(\sigma^{-1}) = \mathsf{Des}(\pi^{-1})$, it follows that $\mathsf{Des}(\sigma)\sigma = \mathsf{Des}(\pi)\pi$, hence $\mathsf{des}(\sigma) = k$ and $\sigma \in \mathcal{S}_n(\underline{k})$. In other words, $\underline{k}\sigma = \underline{k}\pi$, and both σ and π are decreasing on $\underline{k+1}$ and increasing on $\underline{n} \setminus \underline{k}$, by (i). This shows $\pi = \sigma$. Conversely, for any $D \subseteq \underline{n-1}$, a permutation $\sigma \in \mathcal{S}_n$ such that $\mathsf{Des}(\sigma) = \underline{|D|}$ and $\underline{|D|}\sigma = D+1$ may be easily constructed. $\square$

Note that, by 11.7(i) and the definition of ω_n,

$$\omega_n = \sum_{k=0}^{n-1} (-1)^k \Delta^{\underline{k}} = \sum_{\pi \in \mathcal{V}_n} (-1)^{1\pi^{-1}-1} \pi. \tag{11.1}$$

This is the reason for the interest in valley permutations in this setting.

We proceed to a characterisation of rim hooks.

11.8 Definition and Remark. Each $(r,s) \in \mathbb{Z} \times \mathbb{Z}$ has two predecessors, $(r-1,s)$ and $(r,s-1)$, and two successors, $(r+1,s)$ and $(r,s+1)$, in $(\mathbb{Z} \times \mathbb{Z}, \leq_{\mathbb{Z}\times\mathbb{Z}})$ all four of which are considered as neighbours of (r,s) in what follows. This defines the structure of a simple undirected graph on $\mathbb{Z} \times \mathbb{Z}$. As usual, a subset S of $\mathbb{Z} \times \mathbb{Z}$ is said to be *connected* if for any $x, y \in S$ there exists a path $(x_0, \ldots, x_m)$ in S such that $x_0 = x$, $x_m = y$ and x_{i-1} is a neighbour of x_i for all $i \in \underline{m}$.

Note that, if F and G are frames and F is isomorphic to G, then F is connected if and only if G is connected. For, if $\varphi : F \to G$ is an isomorphism, then x and y are neighbours in F if and only if $x\varphi$ and $y\varphi$ are neighbours in G, for all $x, y \in F$, since F and G are convex.

Convexity also leads to a characterisation of connected frames.

11.9 Proposition. *Let F be a frame, then the following conditions are equivalent:*

(i) *F is connected;*

(ii) *if $(r,s) \in F$ and (r,s) is not maximal in $(F, \to)$, then $(r-1,s) \in F$ or $(r,s+1) \in F$;*

(iii) *if F is isomorphic to the coupling of a frame G with a frame H, then $G = \emptyset$ or $H = \emptyset$.*

Proof. Let G and H be nonempty frames, then the coupling of G with H is clearly not connected. Therefore (i) implies (iii).

Assume (iii) and let $(r,s) \in F$ such that $(r-1,s)$ and $(r,s+1)$ are not contained in F. Then F is isomorphic to the coupling of

$$G := \{x \in F \mid x \to (r,s)\}$$

with $H := F \backslash G$, since F is convex. Now $(r,s) \in G$, hence (iii) implies $H = \emptyset$. It follows that (r,s) is maximal in $(F, \to)$, thus (ii).

To see that (ii) implies (i), let $x, y \in F$. We may assume that $x \to y$. If, in addition, $x \leq_{\mathbb{Z}\times\mathbb{Z}} y$ or $y \leq_{\mathbb{Z}\times\mathbb{Z}} x$, then any element $z \in (\mathbb{Z}\times\mathbb{Z}, \leq_{\mathbb{Z}\times\mathbb{Z}})$ between x and y is contained in F. In particular, there is a path from x to y in F as desired.

Assume now that x and y are incomparable with respect to $\leq_{\mathbb{Z}\times\mathbb{Z}}$ and let $x = (r,s)$. Then, by (ii), $(r-1,s)$ or $(r,s+1)$ lies in F, since $x \to y$ and $x \neq y$. A simple induction on the distance of x and y in the graph $\mathbb{Z}\times\mathbb{Z}$ completes the proof. □

11.10 Proposition. *Let F be a frame, then F is a rim hook if and only if the following conditions hold:*

(i) *F is connected;*

(ii) *any ideal of $(F, \to)$ is a convex subset of $(F, \leq_{\mathbb{Z}\times\mathbb{Z}})$.*

Proof. Conditions (i) and (ii) are almost part of the definition of a rim hook.

Conversely, assuming (i) and (ii), we use induction on $|F|$ to show that F is a rim hook. If $|F| \leq 1$, then F is a rim hook anyway.

Suppose that $|F| > 1$. Let (r,s) be the largest element of the chain $(F, \to)$, and put $F' := F \backslash \{(r,s)\}$. Any ideal of $(F', \to)$ is also an ideal of $(F, \to)$, hence convex with respect to $\leq_{\mathbb{Z}\times\mathbb{Z}}$. In particular, F' is a convex subset of $(\mathbb{Z}\times\mathbb{Z}, \leq_{\mathbb{Z}\times\mathbb{Z}})$, that is to say, a frame.

Furthermore, 11.9(ii) implies that F' is connected. Hence F' is a rim hook, by the induction hypothesis. Let (r', s') be the largest element of $(F', \to)$, then $(r,s) = (r', s'+1)$, or $(r,s) = (r'-1, s')$, again by 11.9(ii). This shows that F is also a rim hook. □

This chapter concludes with the

Proof of 11.3. We first prove the claim for $\zeta^F(C_n)$. By definition, $\zeta^F(C_n) = c(\mathrm{Z}^F)(C_n) = (\mathrm{Z}^F, \omega_n)_{\mathcal{P}}$. If $|F| \neq n$, this implies $\zeta^F(C_n) = 0$. Let $|F| = n$. If F is not connected, then there exist nonempty frames G, H such that $\mathrm{Z}^F = \mathrm{Z}^G \star \mathrm{Z}^H$, by 11.9 and 6.9. Hence primitivity of ω_n implies

$$\zeta^F(C_n) = (\mathrm{Z}^G \star \mathrm{Z}^H, \omega_n)_{\mathcal{P}}$$

$$= (\mathrm{Z}^G \otimes \mathrm{Z}^H, \omega_n\!\downarrow)_{\mathcal{P}\otimes\mathcal{P}}$$
$$= (\mathrm{Z}^G \otimes \mathrm{Z}^H, \omega_n \otimes \emptyset + \emptyset \otimes \omega_n)_{\mathcal{P}\otimes\mathcal{P}}$$
$$= 0.$$

Assume now that $(\mathrm{Z}^F, \omega_n)_{\mathcal{P}} \neq 0$, so F is connected. Besides, there is a valley permutation $\pi \in \mathcal{V}_n$, by (11.1), such that $(\mathrm{Z}^F, \pi)_{\mathcal{P}} \neq 0$, that is to say, $\pi^{-1} \in \mathsf{SYT}^F$. The mapping

$$\alpha := \iota_F^{-1}\pi^{-1} : (F, \leq_F) \to (\underline{n}_|, \leq)$$

is thus a monotone bijection. If I is an ideal of $(F, \to)$ of order j, then $I = \underline{j}_| \iota_F$, by the definition of ι_F, which implies that $I\alpha = \underline{j}_| \iota_F \alpha = \underline{j}_| \pi^{-1}$ is an interval in $(\underline{n}_|, \leq)$, by 11.7(ii). In particular, $I\alpha$ is convex in $(\underline{n}_|, \leq)$, hence I is convex in $(F, \leq_{\mathbb{Z}\times\mathbb{Z}})$. Applying 11.10, shows that F is indeed a rim hook.

Let $r = r_1.\dots.r_l$ denote the type of F and set $D := \{r_1, r_1 + r_2, \dots, r_1 + \dots + r_{l-1}\} \subseteq \underline{n-1}_|$. There exists a unique valley permutation $\sigma \in \mathcal{V}_n$ such that $\mathsf{Des}(\sigma^{-1}) = D$, by 11.7. This implies

$$(\mathrm{Z}^F, \omega_n)_{\mathcal{P}} = \sum_{\pi \in \mathcal{V}_n} (-1)^{1\pi^{-1}-1} (\mathrm{Z}^F, \pi)_{\mathcal{P}} = (-1)^{1\sigma^{-1}-1} = (-1)^{\mathsf{leg}(F)},$$

and thus proves the claim in the special case $q = \varnothing$.

The general case may now be derived from the restriction rule 5.10:

$$\zeta^F(C_{q.n}) = (\mathrm{Z}^F, \omega_q \star \omega_n)_{\mathcal{P}}$$
$$= (\mathrm{Z}^F\!\downarrow, \omega_q \otimes \omega_n)_{\mathcal{P}\otimes\mathcal{P}}$$
$$= \sum_{I \trianglelefteq F} (\mathrm{Z}^I \otimes \mathrm{Z}^{F\setminus I}, \omega_q \otimes \omega_n)_{\mathcal{P}\otimes\mathcal{P}}$$
$$= \sum_{I \trianglelefteq F} (\mathrm{Z}^{F\setminus I}, \omega_n)_{\mathcal{P}} (\mathrm{Z}^I, \omega_q)_{\mathcal{P}}$$
$$= \sum_{I} (-1)^{\mathsf{leg}(F\setminus I)} \zeta^I(C_q),$$

where, as asserted, the latter sum is taken over all $I \trianglelefteq F$ such that $F\setminus I$ is a rim hook of order n. □

Chapter 12

Young Characters

Historically, the Young characters ξ^q of $\mathcal{S}_n$ — induced by the trivial characters of the Young subgroups $\mathcal{S}_q$ — play a major role in the representation theory of $\mathcal{S}_n$. They are easy to handle, thanks to their definition, and yet closely connected to the irreducible characters of $\mathcal{S}_n$.

In the course of the noncommutative approach presented so far, the use of Young characters was reduced to a standard argument in the proof of 10.2 showing that the class functions ζ^p are actually characters. As a consequence, many classical results on Young characters may now be derived at once using their noncommutative counterparts Ξ^q.

The first result is

12.1 Young's Rule. *For all $m, n \in \mathbb{N}$ and all $p \vdash m$,*

$$\zeta^p \bullet \xi^n = \sum_r \zeta^r,$$

where the sum is taken over all partitions r of $m+n$ such that $p \subseteq r$ and $\mathsf{F}(r \backslash p)$ is a horizontal strip (that is, isomorphic to $\mathsf{HS}(q)$ for some $q \in \mathbb{N}^*$, see 6.15).

Proof. Let $r \vdash m+n$, then by a noncommutative computation,

$$\begin{aligned}
(\zeta^p \bullet \xi^n, \zeta^r)_{\mathcal{C}} &= (\mathrm{Z}^p \star \Xi^n, \mathrm{Z}^r)_{\mathcal{P}} \\
&= (\mathrm{Z}^p \otimes \Xi^n, \mathrm{Z}^r\!\downarrow)_{\mathcal{P}\otimes\mathcal{P}} \\
&= \sum_{q \subseteq r} (\mathrm{Z}^p, \mathrm{Z}^q)_{\mathcal{P}} (\Xi^n, \mathrm{Z}^{r\backslash q})_{\mathcal{P}}
\end{aligned}$$

$$= \begin{cases} (\Xi^n, Z^{r\backslash p})_{\mathcal{P}}, & \text{if } p \subseteq r \\ 0 & \text{otherwise.} \end{cases}$$

Furthermore, if $p \subseteq r$ then $(\Xi^n, Z^{r\backslash p})_{\mathcal{P}}$ is one or zero according as $\mathsf{id}_n \in \mathsf{SYT}^{\mathsf{F}(r\backslash p)}$ or not. By 6.17, $\mathsf{id}_n \in \mathsf{SYT}^{\mathsf{F}(r\backslash p)}$ if and only if $\mathsf{F}(r\backslash p)$ is a horizontal strip. □

Note that the case $n = 1$ is the *Branching Rule* of Chapter 11.

12.2 Definition. Let $n \in \mathbb{N}_0$, then the matrix $K_n := (k_{qp})_{q,p\vdash n}$, defined by

$$\xi^q = \sum_{p\vdash n} k_{qp}\zeta^p,$$

for all $q \vdash n$, is called the *Kostka matrix* of $\mathcal{S}_n$ and the number k_{qp} the *Kostka number* corresponding to p and q.

As a consequence of 10.1, there is the following result.

12.3 Proposition. *Let $n \in \mathbb{N}$, then the Kostka matrix K_n is upper unitriangular with respect to lexicographically increasing order of row and column indices.*

In particular, $\{\xi^p \,|\, p \vdash n\}$ is a linear basis of $\mathcal{C}\ell_K(\mathcal{S}_n)$.

Though a combinatorial description of the Kostka numbers may be obtained inductively from Young's Rule, observe that another application of the Main Theorem implies directly for all $q, p \vdash n$ that

$$k_{qp} = (\xi^q, \zeta^p)_{\mathcal{C}} = (\Xi^q, Z^p)_{\mathcal{P}} = \mathsf{syt}^p(\mathcal{S}^q) = |(\mathcal{S}^q)^{-1} \cap \mathsf{SYT}^p|.$$

The Kostka number k_{qp} is thus equal to the number of standard Young tableaux π of shape p such that $\pi^{-1} \in \mathcal{S}^q$. This combinatorial description is readily seen to coincide with the one commonly given in the literature using the notion of standard tableau of shape p and content q. We will demonstrate this now for the sake of completeness.

12.4 Definition. Let $n \in \mathbb{N}_0$ and F be a frame of order n, then any word $w = w_1.\ldots.w_n \in \mathbb{N}^*$ is a *standard tableau of shape F* if the mapping $\alpha : F \to \mathbb{N}$, defined by $i\iota_F\alpha = w_i$ for all $i \in \underline{n}$, has the following properties:

(i) if $x \leq_{\mathbb{Z}\times\mathbb{Z}} y$ then $x\alpha \leq y\alpha$;

(ii) if $x <_{\mathbb{Z}\times\mathbb{Z}} y$ and $y \to x$ then $x\alpha < y\alpha$, for all $x, y \in F$. Denote by ST^F the set of all standard tableaux of shape F.

12.5 Example. As usual, the pictorial description helps to clarify the definition. Let F be a frame and $x \in F$. Insert a diagonal of negative slope into the cells $y \in F\backslash\{x\}$ such that $x \leq_{\mathbb{Z}\times\mathbb{Z}} y$, and a diagonal of positive slope into the cells $y \in F\backslash\{x\}$ such that $y \to x$ to obtain

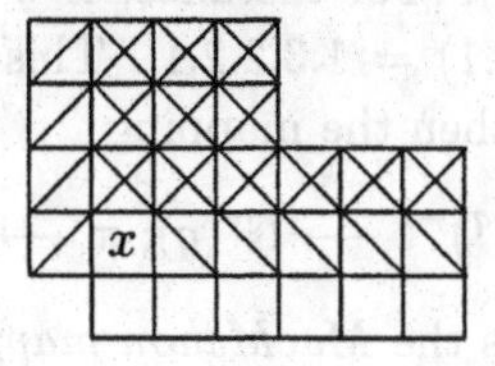

According to condition (ii), $x\alpha < y\alpha$ for all cells $y \in F$ marked by a cross, while condition (i) means that $x\alpha \leq y\alpha$ for all cells y marked by a diagonal of negative slope. Combining both properties, the word $w = w_1.\dots.w_n$, defined by $w_i = i\iota_F\alpha$ for all $i \in \underline{n}$, is a standard tableau of shape F if and only if α is weakly increasing in the rows and strictly decreasing in the columns of F.

For instance, the mapping $\alpha : \mathsf{F}(5.4.3.1\backslash 2.2.1) \to \mathbb{N}$, illustrated by

2				
	4	4		
		2	4	
		1	3	3

corresponds to the standard tableau $w = 2.4.4.2.4.1.3.3$ of shape $\mathsf{F}(5.4.3.1\backslash 2.2.1)$.

12.6 Definition. Let $n \in \mathbb{N}_0$ and $q = q_1.\dots.q_k \models n$. Denote by $\mathbb{N}^*(q)$ the set of all words r of length n in $\mathbb{N}^*$ such that the multiplicity of the letter j in r is q_j, for all $j \in \underline{k}$. The elements of $\mathbb{N}^*(q)$ are called *words of content* q. For instance, the words $w = 2.1.2.3.1$ and $\tilde{w} = 1.3.2.2.1$ both have content 2.2.1 and the standard tableau mentioned in 12.5 above has content 1.2.2.3. For all $T \subseteq \mathbb{N}^*$, let

$$T(q) := T \cap \mathbb{N}^*(q)$$

be the set of all words of content q in T. Furthermore, if F is a frame, then we set

$$\mathsf{st}^F(q) := |\mathsf{ST}^F(q)| \quad \text{and} \quad \mathsf{st}^p(q) := |\mathsf{ST}^p(q)|,$$

for all $q, p \in \mathbb{N}^*$ such that p is a partition.

12.7 Definition and Remark. Let $n \in \mathbb{N}_0$ and $q = q_1.\dots.q_k \models n$. Using the Polya action defined in 5.4, the symmetric group $\mathcal{S}_n$ acts on $\mathbb{N}^*(q)$ from the left via

$$(\pi, w) \longmapsto \pi w$$

for all $\pi \in \mathcal{S}_n$, $w \in \mathbb{N}^*(q)$. For instance, if $\pi = 2\,4\,1\,3\,5 \in \mathcal{S}_5$ (viewed as a word), then $\pi(2.1.2.3.1) = 1.3.2.2.1$. This action is transitive. Let $w_q := 1^{q_1}.\dots.k^{q_k} \in \mathbb{N}^*(q)$, then the mapping

$$\mathsf{M} : (\mathcal{S}^q)^{-1} \longrightarrow \mathbb{N}^*(q),\ \pi \longmapsto \pi w_q$$

is a bijection, referred to as the *MacMahon mapping* [Mac16].

Indeed, let $\pi, \sigma \in (\mathcal{S}^q)^{-1}$, then $\pi w_q = \sigma w_q$ implies $\mathcal{S}_q\sigma^{-1} = \mathcal{S}_q\pi^{-1}$ and thus $\sigma = \pi$, since $\mathcal{S}^q$ is a transversal of the right cosets of $\mathcal{S}_q$ in $\mathcal{S}_n$.

12.8 Proposition. *Let F be a frame of order n and $q \models n$, then the MacMahon mapping yields a bijection*

$$\mathsf{SYT}^F \cap (\mathcal{S}^q)^{-1} \longrightarrow \mathsf{ST}^F(q),$$

by restriction.

Proof. Let $q = q_1.\dots.q_k$, $\pi \in (\mathcal{S}^q)^{-1}$ and set $w := \pi\mathsf{M} = \pi w_q$. We need to show that $\pi \in \mathsf{SYT}^F$ if and only if $w \in \mathsf{ST}^F$. Recall that the set partition $P^q = (P_1^q, \dots, P_k^q)$ of $\underline{n}$ consists of the successive blocks of order $q_1, \dots, q_k$ in $\underline{n}$.

Suppose $i, j \in \underline{n}$ such that $i < j$, then there exist indices $u, v \in \underline{k}$ such that $i\pi \in P_u^q$ and $j\pi \in P_v^q$. If $i\pi > j\pi$, then $u \geq v$ and even $u > v$, as $i < j$ and π^{-1} is increasing on P_u^q. It follows that

$$w_i > w_j \iff (w_q)_{i\pi} > (w_q)_{j\pi} \iff u > v \iff i\pi > j\pi.$$

In particular, $\pi \in \mathsf{SYT}^F$ if and only if $w \in \mathsf{ST}^F$ as desired. □

12.9 Theorem. *Let F be a frame, $n \in \mathbb{N}_0$ and $q \models n$, then*

$$(\zeta^F, \xi^q)_{\mathcal{C}} = \mathsf{st}^F(q),$$

which is the number of standard tableaux of shape F and content q. In particular,

$$k_{qp} = \mathsf{st}^p(q),$$

for all $p \vdash n$.

Proof. $(\zeta^F, \xi^q)_{\mathcal{C}} = (\mathrm{Z}^F, \Xi^q)_{\mathcal{P}} = |\mathsf{SYT}^F \cap (\mathcal{S}^q)^{-1}| = |\mathsf{ST}^F(q)|$, by 12.8. □

12.10 Example. Let $n = 6$, $q = 3.2.1$, $p = 4.2 \vdash 6$, then the only two standard tableaux of shape p and content q are 2.3.1.1.1.2 and 2.2.1.1.1.3 and correspond to the pictures

2	3		
1	1	1	2

and

2	2		
1	1	1	3

.

Therefore $(\xi^{3.2.1}, \zeta^{4.2})_{\mathcal{C}} = 2$. The complete Kostka matrix of $\mathcal{S}_6$ is displayed below.

p \ q	1^7	2.1^5	$2^2.1^2$	2^3	3.1^3	$3.2.1$	3^2	4.1^2	4.2	5.1	6
	ζ										
1^7	1	5	9	5	10	16	5	10	9	5	1
2.1^5	0	1	3	2	4	8	3	6	6	4	1
$2^2.1^2$	0	0	1	1	1	4	2	3	4	3	1
2^3	0	0	0	1	0	2	1	1	3	2	1
3.1^3	0	0	0	0	1	2	1	3	3	3	1
$3.2.1$ ξ	0	0	0	0	0	1	1	1	2	2	1
3^2	0	0	0	0	0	0	1	0	1	1	1
4.1^2	0	0	0	0	0	0	0	1	1	2	1
4.2	0	0	0	0	0	0	0	0	1	1	1
5.1	0	0	0	0	0	0	0	0	0	1	1
6	0	0	0	0	0	0	0	0	0	0	1

In general, assuming lexicographically increasing order of row and column indices, the first row of K_n consists of the numbers syt^p, $p \vdash n$, that is to say, the dimensions of the irreducible $\mathcal{S}_n$-modules, according to the decomposition of the regular $\mathcal{S}_n$-character ξ^{1^n}.

Note that indeed $q \leq_{\mathrm{lex}} p$ holds whenever $k_{qp} \neq 0$ in K_6, but the converse is false ($q = 3^2$, $p = 4.1^2$). More generally, 12.9 implies $k_{qp} = 0$ whenever $\ell(p) > \ell(q)$, since the first column of any standard tableau of shape p (entered into the frame $\mathsf{F}(p)$) contains $\ell(p)$ distinct values. A complete characterisation of the non-vanishing entries of the Kostka matrix due to Ruch–Schönhofer is part of 12.18.

We proceed to the classical results on scalar products of Young characters.

12.11 Notation. Let $n \in \mathbb{N}_0$ and $q = q_1.\dots.q_k, r = r_1.\dots.r_l \models n$. Denote by $\mathcal{M}_q^r$ the set of all $l \times k$ matrices $M = (m_{i,j})_{i\in \underline{l}, j\in \underline{k}}$ of nonnegative integers such that

$$\sum_{\nu=1}^{l} m_{\nu,j} = q_j \quad \text{and} \quad \sum_{\mu=1}^{k} m_{i,\mu} = r_i \ ,$$

for all $i \in \underline{l}$, $j \in \underline{k}$. Furthermore, let $\hat{\mathcal{M}}_q^r$ denote the subset of $\mathcal{M}_q^r$ consisting of all matrices of zeros and ones in $\mathcal{M}_q^r$. For instance,

$$M = \begin{pmatrix} 1\,0\,2 \\ 0\,4\,1 \end{pmatrix} \in \mathcal{M}_{1.4.3}^{3.5} \quad \text{and} \quad \hat{M} = \begin{pmatrix} 1\,0\,1 \\ 0\,1\,1 \end{pmatrix} \in \hat{\mathcal{M}}_{1.1.2}^{2.2}.$$

Set $m_q^r := |\mathcal{M}_q^r|$ and $\hat{m}_q^r := |\hat{\mathcal{M}}_q^r|$. Finally, denote the set of all permutations $\pi \in \mathcal{S}_n$ which are *decreasing* on P_i^r for all $i \in \underline{l}$ by $\hat{\mathcal{S}}^r$, and set

$$\hat{\Xi}^r := \Sigma \hat{\mathcal{S}}^r.$$

12.12 Remark. The numbers m_q^r are closely related to the structure constants $m_q^r(s)$ of $\mathcal{D}_n$ occuring in Solomon's multiplication rule (1.1). Namely, for any $q, r, s \models n$, the coefficient $m_q^r(s)$ in (1.1) is equal to the number of matrices $M \in \mathcal{M}_q^r$ such that the word s is obtained by juxtaposing the rows of M from top to bottom and deleting the zeros. For example, for the matrix $M \in \mathcal{M}_{1.4.3}^{3.5}$ considered above, we get $s = 1.2.4.1$. For more details, see B.5 in Appendix B.

12.13 Proposition. $\mathsf{sgn}_n \xi^r = c(\hat{\Xi}^r)$, *for all* $r \in \mathbb{N}^*$.

Proof. Let $n \in \mathbb{N}$, $r = r_1.\dots.r_l \models n$ and $\pi \in \mathcal{S}_n$, then $\pi|_{P_i^r}$ is increasing if and only if $(\pi\rho_n)|_{P_i^r}$ is decreasing, for all $i \in \underline{l}$. This implies $\hat{\Xi}^r = \Xi^r \rho_n$. The claim follows from 9.9 and 10.6(ii). □

Recall from 8.11 that p' denotes the partition conjugate to p, for any partition p.

12.14 Theorem. *Let $n \in \mathbb{N}$ and $q, r \models n$, then*

$$(\xi^q, \xi^r)_{\mathcal{C}} = |\mathcal{S}^q \cap (\mathcal{S}^r)^{-1}| = m_q^r = \sum_{p \vdash n} \mathrm{st}^p(q)\,\mathrm{st}^p(r)$$

and

$$(\xi^q, \mathrm{sgn}_n \xi^r)_{\mathcal{C}} = |\mathcal{S}^q \cap (\hat{\mathcal{S}}^r)^{-1}| = \hat{m}_q^r = \sum_{p \vdash n} \mathrm{st}^p(q)\,\mathrm{st}^{p'}(r).$$

Proof. The third description of each of the scalar products may be obtained from 12.9 and 10.7 via

$$(\xi^q, \xi^r)_{\mathcal{C}} = \sum_{p \vdash n} (\xi^q, \zeta^p)_{\mathcal{C}} (\xi^r, \zeta^p)_{\mathcal{C}}$$

and

$$(\xi^q, \mathrm{sgn}_n \xi^r)_{\mathcal{C}} = \sum_{p \vdash n} (\xi^q, \zeta^p)_{\mathcal{C}} (\mathrm{sgn}_n \xi^r, \zeta^p)_{\mathcal{C}} = \sum_{p \vdash n} (\xi^q, \zeta^p)_{\mathcal{C}} (\xi^r, \zeta^{p'})_{\mathcal{C}},$$

respectively. The noncommutative calculus implies

$$(\xi^q, \xi^r)_{\mathcal{C}} = (\Xi^q, \Xi^r)_{\mathcal{P}} = |\mathcal{S}^q \cap (\mathcal{S}^r)^{-1}|$$

and, by the above proposition,

$$(\xi^q, \mathrm{sgn}_n \xi^r)_{\mathcal{C}} = (\Xi^q, \hat{\Xi}^r)_{\mathcal{P}} = |\mathcal{S}^q \cap (\hat{\mathcal{S}}^r)^{-1}|.$$

Finally, define a mapping from $\mathcal{S}^q$ to $\mathcal{M}_q^r$ as follows. For each $\pi \in \mathcal{S}^q$, let M_π be the $l \times k$ matrix with (i,j)-component $|P_j^q \cap P_i^r \pi^{-1}|$, for all $i \in \underline{k}$, $j \in \underline{l}$, then, indeed, $M_\pi \in \mathcal{M}_q^r$. The proof is complete upon checking that the mapping $\pi \mapsto M_\pi$ yields bijections $\mathcal{S}^q \cap (\mathcal{S}^r)^{-1} \to \mathcal{M}_q^r$ and $\mathcal{S}^q \cap (\hat{\mathcal{S}}^r)^{-1} \to \hat{\mathcal{M}}_q^r$, respectively, by restriction. This is left to the reader. □

The investigation of Young characters is now brought to a close by a derivation of the Ruch–Schönhofer characterisation of the non-vanishing Kostka numbers, and related results.

12.15 Definition and Remark. Let $q = q_1 \dots q_k$ and $p = p_1 \dots p_l$ be compositions of $n \in \mathbb{N}$, then q *is dominated by* p if

$$q_1 + \dots + q_i \leq p_1 + \dots + p_i,$$

for all $i \in \underline{k}$, where $q_i := 0$ or $p_j := 0$ for $i > k$ or $j > l$ if necessary. We write $q \ll p$ in this case.

Note that $q \ll p$ implies $q \leq_{\mathrm{lex}} p$ and $\ell(p) \leq \ell(q)$.

Two helpful observations follow.

12.16 Proposition. *Let $n \in \mathbb{N}$, $p, q = q_1.\dots.q_k \vdash n$ such that $q \ll p$ and $p \neq q$. Then there exist indices $i, j \in \underline{l}$ with $i < j$ such that*

$$\tilde{q} := \begin{cases} q_1.\dots.q_{i-1}.(q_i+1).q_{i+1}.\dots.q_{j-1}.(q_j-1).q_{j+1}.\dots.q_k & \text{if } q_j > 1, \\ q_1.\dots.q_{i-1}.(q_i+1).q_{i+1}.\dots.q_{k-1} & \text{if } q_j = 1, \end{cases}$$

is a partition of n and $q \ll \tilde{q} \ll p$.

Proof. Let $i \in \underline{k}$ be minimal such that $\sum_{\nu=1}^{i} q_i < \sum_{\nu=1}^{i} p_i$, then $i < k$ and $q_{i-1} = p_{i-1} \geq p_i > q_i$ in case $i > 1$. Let $j > i$ be maximal such that $q_j = q_{i+1}$, then $q_j > q_{j+1}$ in case $j < k$. With this choice of i and j, $\tilde{q}$ is a partition of n, and clearly $q \ll \tilde{q}$.

Let $m \in \underline{k}$ and set $s_m(r) := r_1 + \dots + r_m$ for all $r = r_1.\dots.r_l \in \mathbb{N}^*$. Assume that $s_m(p) = s_m(q)$ for some $i < m < j$, and choose m minimal with this property. Then $s_m(p) < n$ implies $m < \ell(p)$, and $s_{m-1}(p) > s_{m-1}(q)$ implies $p_{m+1} \leq p_m < q_m = q_{m+1}$, hence

$$s_{m+1}(p) = s_m(p) + p_{m+1} < s_m(q) + q_{m+1} = s_{m+1}(q),$$

a contradiction. Thus $s_m(p) \geq s_m(q) + 1 = s_m(\tilde{q})$ for all $i \leq m < j$, while $s_m(p) \geq s_m(q) = s_m(\tilde{q})$ holds anyway for $m < i$ and $m \geq j$. This shows $\tilde{q} \ll p$ as asserted. □

12.17 Proposition. *Let $p = p_1.\dots.p_l \in \mathbb{N}^*$ be a partition, then*

$$\sum_{\nu=1}^{j} p_\nu = \sum_{i=1}^{\ell(p')} \min\{p'_i, j\}$$

for all $j \in \underline{l}$.

Proof. Let $l' = \ell(p') = p_1$ and define $\eta_{i\nu}$ to be one or zero according as $i \leq p_\nu$ or not, for all $i \in \underline{l'}$ and $\nu \in \mathbb{N}$, then

$$\sum_{\nu=1}^{j} p_\nu = \sum_{\nu=1}^{j}\sum_{i=1}^{l'} \eta_{i\nu} = \sum_{i=1}^{l'}\sum_{\nu=1}^{j} \eta_{i\nu} = \sum_{i=1}^{l'} \min\{p'_i, j\},$$

for all $j \in \underline{l}$ as asserted. □

12.18 Theorem. *Let $n \in \mathbb{N}$ and $p, q \vdash n$, then the following conditions are equivalent:*

(i) $q \ll p$,

(ii) $(\xi^q, \zeta^p)_{\mathcal{C}} > 0$,

(iii) $\hat{m}_q^{p'} \neq 0$,

(iv) *there exists a standard Young tableau of shape p and content q,*

(v) $\xi^q - \xi^p$ *is a character of* $\mathcal{S}_n$.

The equivalence of (i) and (ii) is known as the Ruch–Schönhofer theorem [RS70; Ruc75], while the equivalence of (i) and (iii) is known as the Gale–Ryser theorem [Gal57; Rys57].

Proof. The equivalence of (ii) and (iv) is immediate from 12.9.

Assume (i) and prove (v) by induction on the lexicographic order of q. If $q = p$, then $\xi^q - \xi^p = 0$.

Suppose $q = q_1.\dots.q_k \neq p$ and choose i, j, $\tilde{q}$ as in 12.16, then $\xi^{\tilde{q}} - \xi^p$ is a character, by induction.

Furthermore, $\chi = \xi^{q_i.q_j} - \xi^{(q_i+1).(q_j-1)}$ (respectively, $\chi = \xi^{q_i.q_j} - \xi^{(q_i+1)}$ if $q_j = 1$) is a character, by Young's Rule 12.1. Choosing $r \in \mathbb{N}^*$ such that $q \approx q_i.q_j.r$, it follows that $\xi^q - \xi^{\tilde{q}} = \chi \bullet \xi^r$ is a character as well, by 3.3. Now (v) follows from

$$\xi^q - \xi^p = (\xi^q - \xi^{\tilde{q}}) + (\xi^{\tilde{q}} - \xi^p).$$

In return, (v) implies $(\xi^q, \zeta^p)_{\mathcal{C}} \geq (\xi^p, \zeta^p)_{\mathcal{C}} = 1$, by 12.3, thus (ii); and (ii) implies

$$\begin{aligned}
\hat{m}_q^{p'} &= (\xi^q, \mathsf{sgn}_n \xi^{p'})_{\mathcal{C}} \\
&= \sum_{r \vdash n} (\xi^q, \zeta^r)_{\mathcal{C}} (\zeta^r, \mathsf{sgn}_n \xi^{p'})_{\mathcal{C}} \\
&\geq (\xi^q, \zeta^p)_{\mathcal{C}} (\mathsf{sgn}_n \zeta^p, \xi^{p'})_{\mathcal{C}} \\
&> 0,
\end{aligned}$$

by 12.14 and 10.7, hence (iii).

Finally, assume (iii) and choose $M = (m_{i,j}) \in \hat{\mathcal{M}}_q^{p'}$, then, by 12.17,

$$\sum_{\nu=1}^{j} q_\nu = \sum_{\nu=1}^{j} \sum_{i=1}^{\ell(p')} m_{i,\nu} = \sum_{i=1}^{\ell(p')} \sum_{\nu=1}^{j} m_{i,\nu} \leq \sum_{i=1}^{\ell(p')} \min\{p'_i, j\} = \sum_{\nu=1}^{j} p_\nu\,,$$

for all $j \in \underline{\ell(q)}$, which shows (i). The proof is complete. □

12.19 Corollary. *Let $n \in \mathbb{N}$ and $p \vdash n$, then ζ^p is the unique common irreducible constituent of ξ^p and $\mathsf{sgn}_n \xi^{p'}$. Furthermore,*

$$(\xi^p, \zeta^p)_{\mathcal{C}} = 1 = (\mathsf{sgn}_n \xi^{p'}, \zeta^p)_{\mathcal{C}}.$$

Proof. On the one hand, by 12.3 and 10.7,

$$(\xi^p, \zeta^p)_{\mathcal{C}} = 1 = (\xi^{p'}, \zeta^{p'})_{\mathcal{C}} = (\xi^{p'}, \mathsf{sgn}_n \zeta^p)_{\mathcal{C}} = (\mathsf{sgn}_n \xi^{p'}, \zeta^p)_{\mathcal{C}}$$

as asserted. On the other hand, for each $q \vdash n$ such that $(\xi^p, \zeta^q)_{\mathcal{C}} \neq 0 \neq (\mathsf{sgn}_n \xi^{p'}, \zeta^q)_{\mathcal{C}}$, it follows that

$$\hat{m}_p^{q'} = (\xi^p, \mathsf{sgn}_n \xi^{q'})_{\mathcal{C}} \geq (\xi^p, \zeta^q)_{\mathcal{C}} > 0$$

and

$$\hat{m}_q^{p'} = (\xi^q, \mathsf{sgn}_n \xi^{p'})_{\mathcal{C}} \geq (\zeta^q, \mathsf{sgn}_n \xi^{p'})_{\mathcal{C}} > 0.$$

By 12.18, this yields $p \ll q$ and $q \ll p$, hence $p = q$. □

Chapter 13

The Littlewood–Richardson Rule

A significant problem of the theory is to specify the structure constants $(\zeta^p \bullet \zeta^q, \zeta^r)_{\mathcal{C}}$ of the algebra $\mathcal{C}$ of class functions. The first combinatorial description of these multiplicities was given by Littlewood and Richardson [LR34]. In the noncommutative approach presented here, the intrinsic description —

13.1 Theorem. *For all partitions $p, q, r \in \mathbb{N}^*$, the multiplicity*

$$(\zeta^p \bullet \zeta^q, \zeta^r)_{\mathcal{C}} = (\mathrm{Z}^p \star \mathrm{Z}^q, \mathrm{Z}^r)_{\mathcal{P}}$$

is equal to the number of standard Young tableaux π of shape r such that π^{-1} is a standard Young tableau of shape U, where U is the coupling of $\mathsf{F}(p)$ with $\mathsf{F}(q)$.

— is also of combinatorial nature, the structure constants being given by the cardinalities of certain sets of permutations.

13.2 Example. Let $p := 3.1$, $q := 4.3.1$ and $r := 5.4.3$. Count the permutations $\pi \in \mathcal{S}_{12}$ such that π and π^{-1} when entered, respectively, in

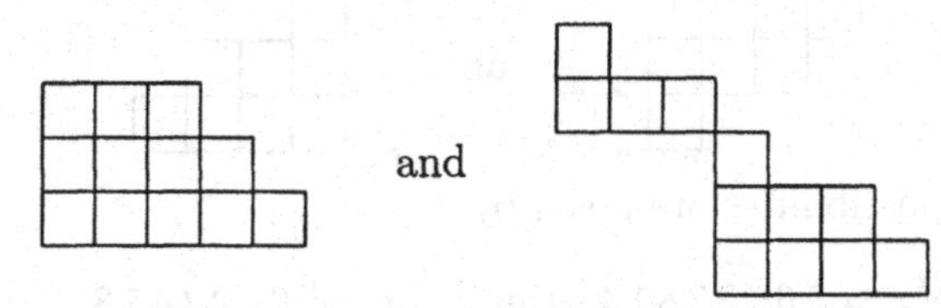

row-wise from top to bottom, are increasing in rows and decreasing in columns. There are precisely two such permutations, namely

$$\pi = 9\,(10)\,(11)\,2\,6\,7\,(12)\,1\,3\,4\,5\,8 \quad \text{and} \quad \sigma = 9\,(10)\,(11)\,2\,6\,7\,8\,1\,3\,4\,5\,(12),$$

hence $(\zeta^{3.1} \bullet \zeta^{4.3.1}, \zeta^{5.4.3})_{\mathcal{C}} = 2$. For instance, entering π and π^{-1}, gives

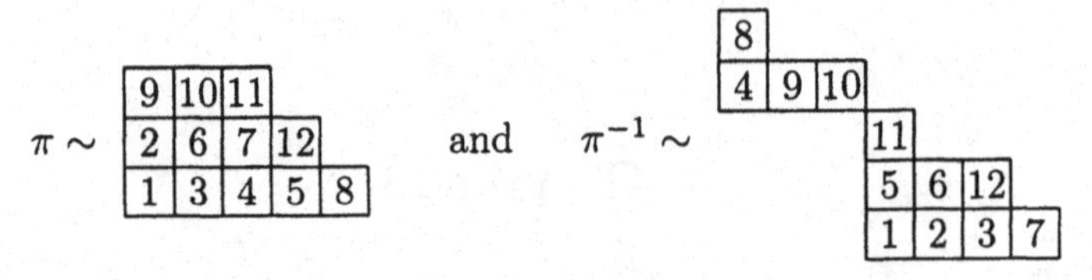

A second description may be obtained as follows.

13.3 Theorem. *For all partitions $p, q, r \in \mathbb{N}^*$, the multiplicity*

$$(\zeta^p \bullet \zeta^q, \zeta^r)_{\mathcal{C}} = \begin{cases} (\mathrm{Z}^q, \mathrm{Z}^{r\backslash p})_{\mathcal{P}} & \text{if } p \subseteq r \\ 0 & \text{otherwise} \end{cases}$$

is equal to the number of standard Young tableaux π of shape q such that π^{-1} is a standard Young tableau of shape $\mathsf{F}(r)\backslash\mathsf{F}(p)$. In particular,

$$(\zeta^p \bullet \zeta^q, \zeta^r)_{\mathcal{C}} = (\zeta^q, \zeta^{r\backslash p})_{\mathcal{C}}.$$

Proof. Using the self-duality 5.14 and applying the restriction rule 6.13,

$$\begin{aligned}(\mathrm{Z}^p \star \mathrm{Z}^q, \mathrm{Z}^r)_{\mathcal{P}} &= (\mathrm{Z}^p \otimes \mathrm{Z}^q, \mathrm{Z}^r\!\downarrow)_{\mathcal{P}\otimes\mathcal{P}} \\ &= \sum_{s \subseteq r} (\mathrm{Z}^p, \mathrm{Z}^s)_{\mathcal{P}} (\mathrm{Z}^q, \mathrm{Z}^{r\backslash s})_{\mathcal{P}} \\ &= \begin{cases} (\mathrm{Z}^q, \mathrm{Z}^{r\backslash p})_{\mathcal{P}}, & \text{if } p \subseteq r \\ 0 & \text{otherwise.} \end{cases}\end{aligned}$$

□

13.4 Example. Consider the partitions $p := 3.1$, $q := 4.3.1$ and $r := 5.4.3$ once more. Look for permutations $\pi \in \mathcal{S}_8$ such that π and π^{-1} may be entered, respectively, in

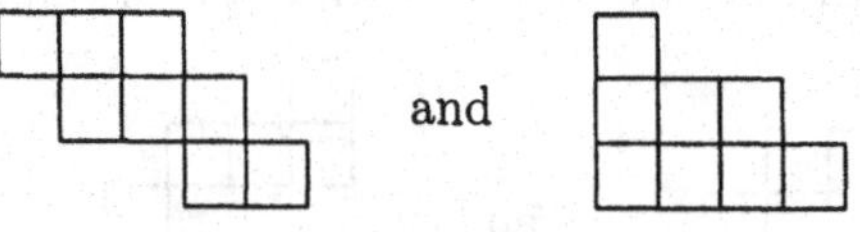

The only two possibilities are given by

$$\pi = 5\,6\,7\,2\,3\,8\,1\,4 \quad \text{and} \quad \sigma = 5\,6\,7\,2\,3\,4\,1\,8,$$

the first of which may be illustrated by:

$$\pi \sim \begin{array}{cccc} 5 & 6 & 7 & \\ & 2 & 3 & 8 \\ & & & 1 \; 4 \end{array} \qquad \text{and} \qquad \pi^{-1} \sim \begin{array}{cccc} 7 & & & \\ 4 & 5 & 8 & \\ 1 & 2 & 3 & 6 \end{array}$$

For the sake of completeness, we note that the second description may be translated into the original Littlewood–Richardson Rule as follows.

Let $q = q_1.\dots.q_k$ be a partition of n and let $\overline{q} := q_k.\dots.q_1$ denote the word in $\mathbb{N}^*$ obtained by reading q backwards. Then $\mathsf{SYT}^q \subseteq \mathcal{S}^{\overline{q}}$, by 6.4 and 6.15.

For any frame F, restricting the MacMahon mapping M defined in 12.7 to

$$L := \mathsf{SYT}^F \cap (\mathsf{SYT}^q)^{-1} \subseteq \mathsf{SYT}^F \cap (\mathcal{S}^{\overline{q}})^{-1}$$

yields an injective mapping $L \to \mathsf{ST}^F(\overline{q})$, by 12.8. If $\pi \in L$ and $w = w_1.\dots.w_n := \pi\mathsf{M}$, the cell $i\pi\iota_{\mathsf{F}(q)}$ is contained in the $(k+1-w_i)$-th row of $\mathsf{F}(q)$, for all $i \in \underline{n}_{\rfloor}$. It follows that, for any $w \in \mathsf{ST}^F(\overline{q})$, there exists an inverse image $\pi \in L$ such that $\pi\mathsf{M} = w$ if and only if

$$\nu_k(w_1.\dots.w_i) \geq \nu_{k-1}(w_1.\dots.w_i) \geq \cdots \geq \nu_1(w_1.\dots.w_i),$$

for all $i \in \underline{k}_{\rfloor}$, where $\nu_j(w_1.\dots.w_i)$ denotes the multiplicity of the letter j in $w_1.\dots.w_i$, for all $j \in \underline{k}_{\rfloor}$.

For instance, for the frame $F := \mathsf{F}(r\backslash p)$ and the permutation $\pi = 5\,6\,7\,2\,3\,8\,1\,4$ considered in Example 13.4,

$$w = \pi(1^1.2^3.3^4) = 3.3.3.2.2.3.1.2,$$

and $\overline{q} = 1.3.4$.

For any such word w, we consider the word $\tilde{w}$ obtained from w by replacing the letter i by $k+1-i$, for all $i \in \underline{k}_{\rfloor}$. Then $\tilde{w}$ has content q and is a so-called *lattice permutation*, that is to say, it has the characteristic property

$$\nu_1(\tilde{w}_1.\dots.\tilde{w}_i) \geq \nu_2(\tilde{w}_1.\dots.\tilde{w}_i) \geq \cdots \geq \nu_k(\tilde{w}_1.\dots.\tilde{w}_i),$$

for all $i \in \underline{k}_{\rfloor}$.

The same example gives the lattice permutation

$$\tilde{w} = 1.1.1.2.2.1.3.2$$

of content $q = 4.3.1$.

Finally, rotate F by 180 degrees to obtain the frame S, so $w \in \mathsf{ST}^F(\overline{q})$ if and only if $\tilde{w}$, now entered row-wise from bottom right to top left, is increasing in the rows and strictly decreasing in the columns of S.

In the example, the rotated frame S is

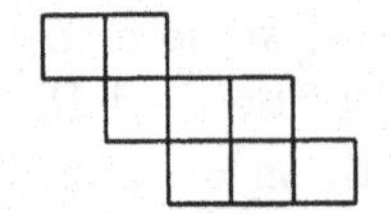

and indeed, the lattice permutation $\tilde{w} = 1.1.1.2.2.1.3.2$, entered from bottom right to top left, is increasing in rows and strictly decreasing in columns.

But

$$\mathrm{Z}^S = \varrho_n \mathrm{Z}^F \varrho_n$$

implies $\zeta^S = \zeta^F$, by 9.9, which proves the

13.5 Littlewood–Richardson Rule. *For all partitions p, q, r, the multiplicity*

$$(\zeta^p \bullet \zeta^q, \zeta^r)_C$$

is equal to the number of lattice permutations of content q in $\mathbb{N}^*$ which, once entered in $\mathsf{F}(r\backslash p)$ row-wise from bottom right to top left, are increasing in the rows and strictly decreasing in the columns of $\mathsf{F}(r\backslash p)$.

It seems worth mentioning that, in the case of a horizontal strip $\mathsf{F}(r\backslash s) = \mathsf{HS}(q)$, Theorem 13.3 combined with the Littlewood–Richardson Rule leads to another description of the Kostka numbers

$$k_{q,p} = (\xi^q, \zeta^p)_{\mathcal{C}} = (\zeta^{r\backslash s}, \zeta^p)_{\mathcal{C}} = (\zeta^r, \zeta^p \bullet \zeta^s)_{\mathcal{C}}$$

in terms of lattice permutations.

This chapter concludes with the following

13.6 Remark. Consider, more generally, four partitions p, q, r and s, then

$$(\zeta^{r\backslash p}, \zeta^{s\backslash q})_{\mathcal{C}} = (\mathrm{Z}^{r\backslash p}, \mathrm{Z}^{s\backslash q})_{\mathcal{P}} = |\mathsf{SYT}^{r\backslash p} \cap (\mathsf{SYT}^{s\backslash q})^{-1}|$$

is equal to the number of standard Young tableaux π of shape $\mathsf{F}(r\backslash p)$ such that π^{-1} is a standard Young tableau of shape $\mathsf{F}(s\backslash q)$. It is readily seen that this description of the scalar product of two skew characters coincides with the one given by Zelevinsky in terms of *pictures* [Zel81a].

Chapter 14

Foulkes Characters and Descent Characters

A brief demonstration of the use of noncommutative character theory with a view to combinatorial applications follows. Consider the *Eulerian numbers*, defined for all $k, n \in \mathbb{N}_0$ by

$$e^{n,k} := \#\{\, \pi \in \mathcal{S}_n \,|\, \mathsf{des}(\pi) = k \,\}.$$

Foulkes suggested looking for proper $\mathcal{S}_n$-characters $\eta^{n,k}$ of degree $e^{n,k}$ and applying character theoretical techniques to study these numbers [Fou80]. In the noncommutative setting, the observation 10.4 is a perfect tool for providing characters with a given degree. For the Eulerian numbers, define

$$E^{n,k} := \{\, \pi \in \mathcal{S}_n \,|\, \mathsf{des}(\pi) = k \,\}$$

and

$$\mathsf{H}^{n,k} := \Sigma E^{n,k} \quad \text{and} \quad \eta^{n,k} := \mathsf{c}(\mathsf{H}^{n,k}),$$

for all $n, k \in \mathbb{N}_0$. Observe $\mathsf{H}^{n,k} \in \mathcal{Q}$, since any two coplactic neighbours have the same descent set, and, in particular, the same number of descents. Applying 10.4 yields:

14.1 Theorem. *Let $n \in \mathbb{N}$, then for all $k \in \mathbb{N}_0$, $\eta^{n,k}$ is a character of $\mathcal{S}_n$ of degree $e^{n,k}$, called the* Foulkes character *corresponding to n and k [Ker91], and*

$$\eta^{n,k} = \sum_{p \vdash n} \mathsf{syt}^p(E^{n,k})\, \zeta^p = \sum_{p \vdash n} |\mathsf{SYT}^p \cap (E^{n,k})^{-1}|\, \zeta^p.$$

Furthermore, the sum $\sum_{k=0}^{n-1} \eta^{n,k}$ is the regular $\mathcal{S}_n$-character, and $\eta^{n,k} = 0$

whenever $k \geq n$. In particular,

$$e^{n,k} = \sum_{p \vdash n} \mathsf{syt}^p(E^{n,k})\, \mathsf{syt}^p = \sum_{p \vdash n} |\mathsf{SYT}^p \cap (E^{n,k})^{-1}|\, \mathsf{syt}^p$$

and

$$\sum_{k=0}^{n-1} e^{n,k} = n!\,.$$

A nice illustrative description of the decomposition numbers

$$\mathsf{syt}^p(E^{n,k}) = |\mathsf{SYT}^p \cap (E^{n,k})^{-1}|$$

of the Foulkes characters is based on the following observation.

14.2 Proposition. *Let $n \in \mathbb{N}_0$, $p \vdash n$ and $\pi \in \mathsf{SYT}^p$, and put $\alpha := \iota_{\mathsf{F}(p)}^{-1}\pi$, then for all $i \in \underline{n-1}$,*

$$i \in \mathsf{Des}(\pi^{-1}) \iff i\alpha^{-1}P_1 < (i+1)\alpha^{-1}P_1\,,$$

where $P_1 : \mathbb{Z} \times \mathbb{Z} \to \mathbb{Z}$, $(x,y) \longmapsto x$ denotes the projection onto the first component.

In other words, i is a descent of π^{-1} if and only if i stands strictly below $(i+1)$ in $\mathsf{F}(p)$, once π is entered in $\mathsf{F}(p)$ row-wise from top to bottom.

Proof. The definitions of ι_F and α imply $i\pi^{-1} > (i+1)\pi^{-1}$ if and only if

$$(y_1, y_2) := (i+1)\alpha^{-1} = (i+1)\pi^{-1}\iota_F \longrightarrow i\pi^{-1}\iota_F = i\alpha^{-1} =: (x_1, x_2),$$

for arbitrary π. The latter is equivalent to $x_1 < y_1$, or $x_1 = y_1$ and $x_2 > y_2$, by the definition of $\to$. But $x_1 = y_1$ and $x_2 > y_2$ would imply that $y \leq_{\mathbb{Z}\times\mathbb{Z}} x$ and thus $i+1 = y\alpha < x\alpha = i$, since $\pi \in \mathsf{SYT}^p$. It follows that this case never occurs. □

14.3 Example. For the standard Young tableau $\pi = 6\,2\,5\,1\,3\,4$ of shape $p = 3.2.1$,

$$\alpha \sim \begin{array}{|c|c|c|}\hline 6 & \multicolumn{2}{c}{} \\ \cline{1-2} 2 & 5 & \multicolumn{1}{c}{} \\ \hline 1 & 3 & 4 \\ \hline \end{array}\,.$$

The elements $i \in \underline{5}$ such that i stands strictly below $i+1$ are printed in boldfaced type, and $\mathsf{Des}(\pi^{-1}) = \mathsf{Des}(4\,2\,5\,6\,3\,1) = \{1,4,5\}$. The complete list of standard Young tableaux π of shape 3.2.1 such that $\mathsf{des}(\pi^{-1}) = 3$ is given by

$$\begin{array}{lll} 3 & & \\ 2 & 6 & \\ 1 & 4 & 5 \end{array} \qquad \begin{array}{lll} 6 & & \\ 3 & 5 & \\ 1 & 2 & 4 \end{array} \qquad \begin{array}{lll} 4 & & \\ 3 & 6 & \\ 1 & 2 & 5 \end{array} \qquad \begin{array}{lll} 6 & & \\ 2 & 4 & \\ 1 & 3 & 5 \end{array}$$

$$\begin{array}{lll} 4 & & \\ 2 & 6 & \\ 1 & 3 & 5 \end{array} \qquad \begin{array}{lll} 5 & & \\ 2 & 4 & \\ 1 & 3 & 6 \end{array} \qquad \begin{array}{lll} 3 & & \\ 2 & 5 & \\ 1 & 4 & 6 \end{array}$$

together with the one already mentioned. In particular, 14.1 implies $(\eta^{6,3}, \zeta^{3.2.1})_{\mathcal{C}} = 8$.

Foulkes characters have a natural decomposition into descent characters, which are defined as follows.

14.4 Notation. Mapping

$$r \longmapsto D(r) := \{r_1, r_1 + r_2, \ldots, r_1 + \cdots r_{l-1}\} \subseteq \underline{n-1}_|,$$

defines a bijection $\{\, r \mid r \models n \,\} \to \{\, T \mid T \subseteq \underline{n-1}_| \}$, and $|D(r)| = \ell(r) - 1$, for all $r \models n$.

14.5 Definition and Remark. Let $r \in \mathbb{N}^*$ and set $D := D(r)$. Recall that the rim hook $\mathsf{RH}(r)$ of type r has been defined in 11.2. We set

$$\Delta^r := \mathsf{Z}^{\mathsf{RH}(r)} = \sum_{\mathsf{Des}(\pi)=D} \pi = \Sigma\mathcal{S}_n(D)$$

and

$$\delta^r := c(\Delta^r).$$

δ^r is the *descent character* corresponding to r. Since $E^{n,k}$ is the disjoint union of the descent classes $\mathcal{S}_n(D)$ ($D \subseteq \underline{n-1}_|$, $|D| = k$),

$$\mathsf{H}^{n,k} = \sum_{\substack{r \models n \\ \ell(r)=k+1}} \Delta^r \quad \text{and} \quad \eta^{n,k} = \sum_{\substack{r \models n \\ \ell(r)=k+1}} \delta^r$$

for all $n, k \in \mathbb{N}_0$. In particular, the sum of all descent characters δ^r, $r \models n$, is the regular $\mathcal{S}_n$-character.

Another application of 10.4 is:

14.6 Theorem. *Let $q \models n$ and set $D := D(q)$. Then δ^q is a character of $\mathcal{S}_n$ of degree $|\mathcal{S}_n(D)|$. For all $p \vdash n$, the multiplicity of the irreducible* *character ζ^p in δ^q is given by*

$$(\delta^q, \zeta^p)_{\mathcal{C}} = (\Delta^q, \mathrm{Z}^p)_{\mathcal{P}} = \mathsf{syt}^p(\mathcal{S}_n(D)) = |\mathcal{S}_n(D)^{-1} \cap \mathsf{SYT}^p|,$$

which equals the number of standard Young tableaux π of shape p such that $\mathsf{Des}(\pi^{-1}) = D$.

14.7 Example. The illustrations of all standard Young tableaux π of shape 3.2.1 such that $\mathsf{des}(\pi^{-1}) = 3$ given in Example 14.3 yields

$$\begin{aligned}
(\delta^{3.1.1.1}, \zeta^{3.2.1})_{\mathcal{C}} &= 0 &&= (\delta^{1.1.1.3}, \zeta^{3.2.1})_{\mathcal{C}}\,, \\
(\delta^{1.3.1.1}, \zeta^{3.2.1})_{\mathcal{C}} &= 1 &&= (\delta^{1.1.3.1}, \zeta^{3.2.1})_{\mathcal{C}}\,, \\
(\delta^{2.2.1.1}, \zeta^{3.2.1})_{\mathcal{C}} &= 1 &&= (\delta^{1.1.2.2}, \zeta^{3.2.1})_{\mathcal{C}}\,, \\
(\delta^{2.1.2.1}, \zeta^{3.2.1})_{\mathcal{C}} &= 1 &&= (\delta^{1.2.1.2}, \zeta^{3.2.1})_{\mathcal{C}}\,, \\
(\delta^{2.1.1.2}, \zeta^{3.2.1})_{\mathcal{C}} &= 0\,, \\
(\delta^{1.2.2.1}, \zeta^{3.2.1})_{\mathcal{C}} &= 2\,.
\end{aligned}$$

The preceding results suggest proceeding to a more detailed analysis of the descent characters.

Recall from 5.9 that

$$q \sqcup r := q_1 . \dots . q_{k-1} . (q_k + r_1) . r_2 . \dots . r_l$$

for all $q = q_1 . \dots . q_k$ and $r = r_1 . \dots . r_l$ in $\mathbb{N}^* \setminus \{\varnothing\}$.

14.8 Theorem. *For all $q, r \in \mathbb{N}^* \setminus \{\varnothing\}$,*

$$\Delta^q \star \Delta^r = \Delta^{q.r} + \Delta^{q \sqcup r}$$

and

$$\delta^q \bullet \delta^r = \delta^{q.r} + \delta^{q \sqcup r}.$$

In particular, the cardinalities

$$f_q := |\mathcal{S}_m(D(q))|,$$

where $m \in \mathbb{N}$ and $q \models m \in \mathbb{N}$, are given by the recursive formula

$$f_{q.n} = \binom{m+n}{m} f_q - f_{q \sqcup n}\,,$$

for all $n \in \mathbb{N}$, and $f_m = 1$, for all $m \in \mathbb{N}$.

Proof. The formula for the product of Δ^q and Δ^r (and hence of δ^q and δ^r) is an immediate consequence of 6.9. Comparison of degrees in the special case $r = n \in \mathbb{N}$ yields the recursive formula. For, $\delta^n = \zeta^n$ is the trivial character of $\mathcal{S}_n$ and

$$\begin{aligned}\deg(\chi \bullet \varphi) &= (\chi \bullet \varphi, \mathsf{ch}_{1^{m+n}})_{\mathcal{C}} \\ &= \sum_{J \subseteq \underline{n+m}} (\chi, \mathsf{ch}_{(1^{m+n})_J})_{\mathcal{C}} (\varphi, \mathsf{ch}_{(1^{m+n})_{CJ}})_{\mathcal{C}} \\ &= \binom{m+n}{m} \deg\chi \deg\varphi,\end{aligned}$$

for all characters χ of $\mathcal{S}_n$, φ of $\mathcal{S}_m$. □

14.9 Corollary. *Let $n \in \mathbb{N}_0$, and denote by T a transversal of the rearrangement classes of the compositions of n, then the set $\{\delta^q \mid q \in \mathsf{T}\}$ is a linear basis of $\mathcal{C}\ell_K(\mathcal{S}_n)$. More precisely, for all $q \in \mathsf{T}$, $p \vdash n$,*

$$(\delta^q, \zeta^p)_{\mathcal{C}} = \begin{cases} 1 & \text{if } p = r, \\ 0 & \text{if } p <_{\text{lex}} r, \end{cases}$$

where r denotes the partition obtained by properly rearranging q.

Proof. Observe that $\mathsf{SYT}^{\mathsf{RH}(q)} \subseteq \mathsf{SYT}^{\mathsf{HS}(q)}$, hence

$$(\Delta^q, Z^p)_{\mathcal{P}} \leq (\Xi^q, Z^p)_{\mathcal{P}},$$

for all $q \in \mathbb{N}^*$ and all partitions p. The unique partition obtained by rearranging q is denoted by r. By 12.3, $(\Delta^q, Z^p)_{\mathcal{P}} \neq 0$ implies that $p \geq_{\text{lex}} r$, since $\xi^q = \xi^r$. Let $q = q_1.\dots.q_k$, and assume that p is a rearrangement of q. Then, putting $q^j := q_1.\dots.q_j$ for all $j \in \underline{k}$ and using induction,

$$\begin{aligned}(\Delta^q, Z^p)_{\mathcal{P}} &= (\Delta^{q^{k-1}} \star \Xi^{q_k}, Z^p)_{\mathcal{P}} - (\Delta^{q^{k-1} \sqcup q_k}, Z^p)_{\mathcal{P}} \\ &= (\Delta^{q^{k-1}} \star \Xi^{q_k}, Z^p)_{\mathcal{P}} \\ &= (\Delta^{q^{k-2}} \star \Xi^{q_{k-1}} \star \Xi^{q_k}, Z^p)_{\mathcal{P}} \\ &\qquad - (\Delta^{q^{k-2} \sqcup q_{k-1}} \star \Xi^{q_k}, Z^p)_{\mathcal{P}} \\ &= (\Delta^{q^{k-2}} \star \Xi^{q_{k-1}} \star \Xi^{q_k}, Z^p)_{\mathcal{P}} \\ &\qquad - (\Delta^{(q^{k-2} \sqcup q_{k-1}).q_k}, Z^p)_{\mathcal{P}} - (\Delta^{q^{k-2} \sqcup q_{k-1} \sqcup q_k}, Z^p)_{\mathcal{P}} \\ &= (\Delta^{q^{k-2}} \star \Xi^{q_{k-1}} \star \Xi^{q_k}, Z^p)_{\mathcal{P}}\end{aligned}$$

$$\begin{aligned} &= \quad \ldots \\ &= (\Xi^{q_1} \star \cdots \star \Xi^{q_k}, Z^p)_{\mathcal{P}} \\ &= (\Xi^q, Z^p)_{\mathcal{P}} \\ &= 1, \end{aligned}$$

as the partitions obtained by rearranging $q^{k-1} \sqcup q_k$, $(q^{k-2} \sqcup q_{k-1}).q_k$, $q^{k-2} \sqcup q_{k-1} \sqcup q_k$ and so on, are lexicographically larger than p. □

14.10 Remark. Let $n \in \mathbb{N}_0$, $q \models n$ and denote the composition of n obtained by reading q backwards by $\overline{q}$, then $\rho_n \Delta^q \rho_n = \Delta^{\overline{q}}$ implies

$$\delta^q = \delta^{\overline{q}},$$

by 9.9. But the equality $\delta^q = \delta^r$ does not necessarily hold for arbitrary q and r contained in the same rearrangement class. For instance, $\delta^{2.1.1} \neq \delta^{1.2.1}$, and the set $\{\delta^4, \delta^{3.1}, \delta^{2.1.1}, \delta^{1.2.1}\}$ is a linear basis of $\mathcal{C}\ell_K(\mathcal{S}_4)$. Other examples can be found in 14.7. In general it is not known (explicitly) when the descent characters δ^q and δ^r coincide. Here is a strange result due to Lehmann [Leh96], which we state without proof.

Denote the n-th power of s with respect to $\sqcup$ by $s \circ n$, for each $s \in \mathbb{N}^* \setminus \{\varnothing\}$ and $n \in \mathbb{N}$, that is $s \circ n := s \sqcup \ldots \sqcup s$ (n factors). Furthermore, let

$$s \circ q := (s \circ q_1).(s \circ q_2). \ldots .(s \circ q_k),$$

for any $q = q_1. \ldots .q_k \in \mathbb{N}^*$, then:

Theorem. *$\delta^q = \delta^r$ if and only if $\delta^{s \circ q} = \delta^{s \circ r}$, for all $q, r, s \in \mathbb{N}^* \setminus \{\varnothing\}$.*

For instance, putting $s := 3.1.2$ and considering the identity $\delta^{2.1} = \delta^{1.2}$,

$$\delta^{3.1.5.1.2.3.1.2} = \delta^{s \circ 2.1} = \delta^{s \circ 1.2} = \delta^{3.1.2.3.1.5.1.2}.$$

Chapter 15

Cyclic Characters and the Free Lie Algebra

In what follows, $n \in \mathbb{N}$ and ε is a primitive n-th root of unity in K.

15.1 Definition and Remarks. Let $\tau = (1\ 2\ \dots\ n) \in \mathcal{S}_n$ denote the standard cycle of length n in $\mathcal{S}_n$, then the cyclic subgroup Z of $\mathcal{S}_n$ generated by τ has order n. The elements

$$\iota_j := \sum_{i=0}^{n-1} \varepsilon^{ij}\tau^i,$$

indexed by $j \in \underline{n-1} \cup \{0\}$, are (up to the factor $\frac{1}{n}$) mutually orthogonal idempotents in the group ring KZ of Z. The corresponding characters

$$\psi_j : Z \to K,\ \tau^i \mapsto \varepsilon^{-ij}$$

of Z are irreducible and of degree one.

The study of the induced characters $(\psi_j)^{\mathcal{S}_n}$, the *cyclic characters* of $\mathcal{S}_n$, closes our investigations. We shall derive a combinatorial description of the decomposition numbers $((\psi_j)^{\mathcal{S}_n}, \zeta^p)_C$ which was discovered by Kraskiewićz and Weyman in 1987 ([KW87], see [KW01]).

The case where $j = 1$ is of particular interest. The character $(\psi_1)^{\mathcal{S}_n}$ provides a link between the Solomon descent algebra and the free Lie algebra which is briefly indicated (but certainly not exploited) at the end of this chapter. Reutenauer's monograph [Reu93] serves as a general reference on the topic (see also [Gar89; GR89; GKL+95; KLT97]).

15.2 Definition. The *major index* of a permutation $\pi \in \mathcal{S}_n$ is the sum of its descents:

$$\mathsf{maj}\,\pi := \sum_{i \in \mathsf{Des}(\pi)} i.$$

Following Klyachko [Kly74], we define

$$\kappa_n(t) := \sum_{\pi \in \mathcal{S}_n} t^{\mathsf{maj}\,\pi}\pi$$

for all $t \in K$ and, in particular, $\kappa_n := \kappa_n(\varepsilon)$. If M_i denotes the sum of all $\pi \in \mathcal{S}_n$ such that $\mathsf{maj}\,\pi \equiv i$ modulo n, for all $i \in \mathbb{N}_0$, then

$$\kappa_n = \sum_{D \subseteq \underline{n-1}|} \varepsilon^{\Sigma D} \Delta^D = \sum_{i=0}^{n-1} \varepsilon^i M_i\,,$$

where $\Sigma D = \sum_{i \in D} i$ for all $D \subseteq \mathbb{N}$.

The first result needed here is due to Klyachko [Kly74]. For convenience, we set $\iota := \iota_1$.

15.3 Theorem. (Klyachko, 1974) *$\iota\kappa_n = n\iota$ and $\kappa_n\iota = n\kappa_n$. In particular, $\kappa_n^2 = n\kappa_n$.*

Proof. Let $\pi \in \mathcal{S}_n$, then $d \in \underline{n}|$ is a *cyclic descent* of π if $d \in \mathsf{Des}(\pi)$ or $d = n$ and $n\pi > 1\pi$. Denote the set of all cyclic descents of π by $\mathsf{cDes}(\pi)$.

Reading modulo n if necessary, observe that $d = n\pi^{-1}$ is a cyclic descent of π, while $d-1$ is not, since $d\pi = n$. Similarly, $d-1$ is a cyclic descent of $\pi\tau$, while d is not, since $(d+1)\pi\tau = n\tau = 1$. All other cyclic descents of π and $\pi\tau$ coincide, that is, $e \neq d-1, d$ is a cyclic descent of π if and only if e is a cyclic descent of $\pi\tau$. It follows that

$$\mathsf{maj}\,(\pi\tau) \equiv \sum_{e \in \mathsf{cDes}(\pi\tau)} j \equiv \sum_{e \in \mathsf{cDes}(\pi)} j - 1 \equiv \mathsf{maj}\,\pi - 1$$

modulo n, for all $\pi \in \mathcal{S}_n$. As a consequence, for all $i \in \underline{n-1}| \cup \{0\}$, we have $\kappa_n\tau^i = \varepsilon^{-i}\kappa_n$, hence $\kappa_n\iota = n\kappa_n$.

To prove the other multiplication rule, observe first that the number $\mathsf{cdes}(\pi)$ of cyclic descents of π is invariant to left multiplication with τ. More precisely, $d \in \underline{n}|$ is a cyclic descent of π if and only if $d-1$ is a cyclic descent of $\tau\pi$ (modulo n), for all $d \in \underline{n}|$. This implies

$$\mathsf{maj}\,(\tau^i\pi) \equiv \mathsf{maj}\,\pi - i\,\mathsf{cdes}(\pi)$$

modulo n, for all $i \in \underline{n-1} \cup \{0\}$, hence

$$\begin{aligned}
\iota\kappa_n &= \sum_{i=0}^{n-1} \sum_{\pi \in \mathcal{S}_n} \varepsilon^{i+\mathsf{maj}\,\pi} \tau^i \pi \\
&= \sum_{\pi \in \mathcal{S}_n} \sum_{i=0}^{n-1} \varepsilon^{i+\mathsf{maj}\,(\tau^{-i}\pi)} \pi \\
&= \sum_{\pi \in \mathcal{S}_n} \varepsilon^{\mathsf{maj}\,\pi} \Big(\sum_{i=0}^{n-1} (\varepsilon^{1-\mathsf{cdes}(\pi)})^i \Big) \pi \\
&= n \sum_{\substack{\pi \in \mathcal{S}_n \\ \mathsf{cdes}(\pi)=1}} \varepsilon^{\mathsf{maj}\,\pi} \pi
\end{aligned}$$

But, for each $d \in \underline{n}$, there is a unique permutation $\pi \in \mathcal{S}_n$ such that $\mathsf{cDes}(\pi) = \{d\}$, namely $\pi = \tau^d$. The above computation may thus be rounded out to $\iota\kappa_n = n\iota$ as asserted. Both formulae imply $\kappa_n^2 = (\frac{1}{n}\kappa_n\iota)\kappa_n = \kappa_n(\frac{1}{n}\iota\kappa_n) = \kappa_n\iota = n\kappa_n$. □

A second crucial property of Klyachko's idempotent was discovered by Leclerc–Scharf–Thibon [LST96].

15.4 Lemma. *Let $i \in \mathbb{N}$ and denote the order of ε^i by d, then*

$$\kappa_n(\varepsilon^i) = \underbrace{\kappa_d(\varepsilon^i) \star \cdots \star \kappa_d(\varepsilon^i)}_{n/d \text{ factors}}.$$

Proof. The proof is done by induction on $m := n/d$. There is nothing to prove for $m = 1$. Let $m > 1$, then inductively and by 5.2,

$$\begin{aligned}
\kappa_d(\varepsilon^i) \star \cdots \star \kappa_d(\varepsilon^i) &= \kappa_d(\varepsilon^i) \star \kappa_{n-d}(\varepsilon^i) \\
&= \sum_{\pi \in \mathcal{S}_d} \sum_{\sigma \in \mathcal{S}_{n-d}} \varepsilon^{i(\mathsf{maj}\,\pi + \mathsf{maj}\,\sigma)} (\pi \# \sigma) \Xi^{d.(n-d)}.
\end{aligned}$$

But $\mathsf{maj}\,(\pi\#\sigma)\nu \equiv \mathsf{maj}\,(\pi\#\sigma) \equiv \mathsf{maj}\,\pi + \mathsf{maj}\,\sigma$ modulo d for all $\pi \in \mathcal{S}_d$, $\sigma \in \mathcal{S}_{n-d}$, $\nu \in \mathcal{S}^{d.(n-d)}$, as is immediate from the definition of $\#$ and $\mathcal{S}^{d.(n-d)}$. The proof is complete upon noting that $\mathcal{S}^{d.(n-d)}$ is a right transversal of $\mathcal{S}_{d.(n-d)}$ in $\mathcal{S}_n$. □

15.5 Corollary. *Let $i \in \mathbb{N}$ and denote the order of ε^i by d, then*

$$c(\kappa_n(\varepsilon^i)) = \mathsf{ch}_{d^{n/d}}.$$

Proof. According to the lemma above, it suffices to prove $c(\kappa_n) = \mathsf{ch}_n$. For this implies $c(\kappa_d(\varepsilon^i)) = \mathsf{ch}_d$, since ε^i is a primitive d-th root of unity, hence

$$c(\kappa_n(\varepsilon^i)) = c(\kappa_d(\varepsilon^i) \star \cdots \star \kappa_d(\varepsilon^i)) = c(\kappa_d(\varepsilon^i)) \bullet \cdots \bullet c(\kappa_d(\varepsilon^i)) = \mathsf{ch}_{d^{n/d}}$$

as asserted.

Choose $a_p \in K$ such that $c(\frac{1}{n}\kappa_n) = \sum_{p \vdash n} a_p \mathsf{char}_p$, then in $\mathcal{C}\ell_K(\mathcal{S}_n)$

$$\sum_{p \vdash n} a_p^2 \mathsf{char}_p = \Big(\sum_{p \vdash n} a_p \mathsf{char}_p\Big)^2 = c\Big((\tfrac{1}{n}\kappa_n)^2\Big) = c(\tfrac{1}{n}\kappa_n) = \sum_{p \vdash n} a_p \mathsf{char}_p ,$$

by Solomon's theorem 1.1 and Klyachko's theorem 15.3. As a consequence, a_p is zero or one, for all $p \vdash n$. Let A denote the set of all partitions p of n with $a_p = 1$, then

$$\begin{aligned}
\tfrac{1}{n} &= (\tfrac{1}{n}\kappa_n, \mathsf{id}_n)_{\mathcal{P}} \\
&= ((\tfrac{1}{n}\kappa_n)^2, \mathsf{id}_n)_{\mathcal{P}} \\
&= (\tfrac{1}{n}\kappa_n, \tfrac{1}{n}\kappa_n)_{\mathcal{P}} \\
&= \Big(\sum_{p \in A} \mathsf{char}_p , \sum_{p \in A} \mathsf{char}_p\Big)_{\mathcal{C}} \\
&= \sum_{p \in A} \frac{1}{p?} .
\end{aligned}$$

Thus, to complete the proof, $a_n = 1$ remains.

Proposition 11.7(iii) gives $(\Delta^D, \omega_n)_{\mathcal{P}} = (-1)^{|D|}$ for all $D \subseteq \underline{n-1}_|$. This implies

$$na_n = c(\kappa_n)(C_n) = (\kappa_n, \omega_n)_{\mathcal{P}} = \sum_{D \subseteq \underline{n-1}_|} (-1)^{|D|} \varepsilon^{\Sigma D} = \prod_{i=1}^{n-1} (1 - \varepsilon^i) = n.$$

□

The preceding results allows us to deduce the main result of this chapter.

15.6 Theorem. (Leclerc, Scharf, Thibon, 1996) *For all $j \in \underline{n-1}_| \cup \{0\}$,*

$$(\psi_j)^{\mathcal{S}_n} = c(M_j).$$

Proof. Let d_i denote the order of ε^i, for all $i \in \mathbb{N}_0$. The cycle type of τ^i is then d_i^{n/d_i}, for all $i \in \mathbb{N}_0$, and the definition of $(\psi_j)^{\mathcal{S}_n}$ implies

$$(\psi_j)^{\mathcal{S}_n}(\pi) = \frac{1}{n}\sum_{i=0}^{n-1} \sum_{\substack{\sigma\in\mathcal{S}_n \\ \sigma^{-1}\pi\sigma=\tau^i}} \varepsilon^{-ij} = \frac{1}{n}\sum_{i=0}^{n-1} \varepsilon^{-ij}\mathsf{ch}_{d_i^{n/d_i}}(\pi)$$

for all $\pi \in \mathcal{S}_n$. Use 15.5 to deduce

$$(\psi_j)^{\mathcal{S}_n} = \frac{1}{n}\sum_{i=0}^{n-1} \varepsilon^{-ij} c(\kappa_n(\varepsilon^i)) = \frac{1}{n}\sum_{k=0}^{n-1}\sum_{i=0}^{n-1}(\varepsilon^{k-j})^i c(M_i) = c(M_j)$$

as asserted. □

In other words, M_j is a noncommutative cyclic character of $\mathcal{S}_n$, for all $j \in \underline{n-1} \cup \{0\}$.

15.7 Corollary. (Kraskiewićz, Weyman, 1987) *Let $p \vdash n$, then*

$$((\psi_j)^{\mathcal{S}_n}, \zeta^p)_{\mathcal{C}} = (M_j, Z^p)_{\mathcal{P}}$$

equals the number of standard Young tableaux π of shape p such that $\mathsf{maj}\,\pi^{-1} \equiv j$ modulo n, for all $j \in \underline{n-1} \cup \{0\}$.

In [LST96], Theorem 15.6 was *derived* from Corollary 15.7. Our proof of Theorem 15.6 follows [JS00], where, more generally, decomposition numbers are given for $\mathcal{S}_n$-characters induced from arbitrary cyclic subgroups of $\mathcal{S}_n$.

Another variant of the Kraskiewićz–Weyman result is the following.

15.8 Theorem. *Let $p \vdash n$ and $i \in \mathbb{N}$, then*

$$\zeta^p(\tau^i) = \sum_{j=0}^{n-1} \varepsilon^{ij}\,\mathsf{syt}_j^p\,,$$

where $\mathsf{syt}_j^p = (M_j, Z^p)_{\mathcal{P}}$ for all $j \in \underline{n-1} \cup \{0\}$.

This result is sometimes attributed to Springer [Spr74].

Proof. Denote the order of τ^i by d, then

$$\zeta^p(\tau^i) = (\zeta^p, \mathsf{ch}_{d^{n/d}})_{\mathcal{C}} = (Z^p, \kappa_n(\varepsilon^i))_{\mathcal{P}} = \sum_{j=0}^{n-1} \varepsilon^{ij}(Z^p, M_j)_{\mathcal{P}},$$

by 15.5. □

15.9 Example. The standard Young tableaux in $\mathcal{S}_4$ are illustrated below.

$$
\begin{array}{|c|c|c|c|}\hline 1 & 2 & 3 & 4 \\ \hline\end{array}
\qquad
\begin{array}{|c|c|c|}\hline 4 \\ \cline{1-1} 1 & 2 & \mathbf{3} \\ \hline\end{array}
\qquad
\begin{array}{|c|c|c|}\hline 3 \\ \cline{1-1} 1 & \mathbf{2} & 4 \\ \hline\end{array}
\qquad
\begin{array}{|c|c|c|}\hline 2 \\ \cline{1-1} \mathbf{1} & 3 & 4 \\ \hline\end{array}
\qquad
\begin{array}{|c|c|}\hline 3 & 4 \\ \hline 1 & \mathbf{2} \\ \hline\end{array}
$$

$$
\begin{array}{|c|}\hline 4 \\ \hline \mathbf{3} \\ \hline \mathbf{2} \\ \hline \mathbf{1} \\ \hline\end{array}
\qquad
\begin{array}{|c|c|}\hline 3 \\ \cline{1-1} \mathbf{2} \\ \cline{1-1} \mathbf{1} & 4 \\ \hline\end{array}
\qquad
\begin{array}{|c|c|}\hline 4 \\ \cline{1-1} 2 \\ \cline{1-1} \mathbf{1} & \mathbf{3} \\ \hline\end{array}
\qquad
\begin{array}{|c|c|}\hline 4 \\ \cline{1-1} \mathbf{3} \\ \cline{1-1} 1 & \mathbf{2} \\ \hline\end{array}
\qquad
\begin{array}{|c|c|}\hline 2 & 4 \\ \hline \mathbf{1} & \mathbf{3} \\ \hline\end{array}
$$

In general, for $p \vdash n$ and $\pi \in \mathsf{SYT}^p$, it follows from 14.2 that $\operatorname{maj}\pi^{-1}$ is the sum of all $i \in \underline{n-1}$ such that i stands strictly below $(i+1)$ in $\mathsf{F}(p)$, once π is entered in the partition frame $\mathsf{F}(p)$ row-wise from top to bottom. These values are printed in boldfaced type in the above illustration. Applying 15.7, we get

$$
\begin{aligned}
(\psi_0)^{\mathcal{S}_4} &= \zeta^4 + \zeta^{2.2} + \zeta^{2.1.1}, \\
(\psi_1)^{\mathcal{S}_4} &= \zeta^{3.1} + \zeta^{2.1.1}, \\
(\psi_2)^{\mathcal{S}_4} &= \zeta^{3.1} + \zeta^{2.2} + \zeta^{1.1.1.1}, \\
(\psi_3)^{\mathcal{S}_4} &= \zeta^{3.1} + \zeta^{2.1.1}.
\end{aligned}
$$

For arbitrary n,

$$\sum_{j=0}^{n-1} (\psi_j)^{\mathcal{S}_n} = \chi_{K\mathcal{S}_n}$$

is the regular $\mathcal{S}_n$-character. Furthermore, $(\psi_j)^{\mathcal{S}_n} = (\psi_k)^{\mathcal{S}_n}$ whenever there is the identity $\gcd(j,n) = \gcd(k,n)$ of greatest common divisors.

Note that the symmetric group $\mathcal{S}_n$ acts on the conjugacy class C_n, by conjugation. The character $(\psi_0)^{\mathcal{S}_n}$ is afforded by this action, since ψ_0 is the trivial character of Z and $Z = C_{\mathcal{S}_n}(\tau)$.

More significantly, the cyclic character $(\psi_1)^{\mathcal{S}_n}$ is intimately linked to the free Lie algebra as we shall explain now, concluding this chapter.

Consider the tensor algebra $T(V)$ over a finite dimensional K-vector space V and denote the homogeneous component of degree n in $T(V)$ by

$$T_n(V) := \underbrace{V \otimes \cdots \otimes V}_{n}.$$

This is an $(\mathcal{S}_n, \mathsf{GL}(V))$-bimodule. The action of $\mathsf{GL}(V)$ from the right is the diagonal action, while for the $\mathcal{S}_n$-action from the left, we have

$$\pi(v_1 \otimes \cdots \otimes v_n) = v_{1\pi} \otimes \cdots \otimes v_{n\pi}$$

for all $\pi \in \mathcal{S}_n$, $v_1, \ldots, v_n \in V$.

At the beginning of the last century Schur studied the structure of $T(V)$ as a $\mathsf{GL}(V)$-module. In his thesis [Sch01] and the famous paper [Sch27], Schur was able to describe the decomposition of $T_n(V)$ into irreducible $\mathsf{GL}(V)$-modules using the irreducible representations of the symmetric group $\mathcal{S}_n$.

The usual Lie bracket $[x, y] := xy - yx$ defines a Lie algebra structure on $T(V)$. A classical result of Witt [Wit37] says that the Lie subalgebra $L(V)$ generated by V is free over any basis of V, and the homogeneous component $L_n(V) := T_n(V) \cap L(V)$ of $L(V)$ is a $\mathsf{GL}(V)$-submodule of $T_n(V)$. Grün showed

$$L_n(V) = \omega_n T_n(V)$$

(see [Mag40]). In 1942, Thrall raised the question how $L_n(V)$ decomposes into irreducible $\mathsf{GL}(V)$-modules [Thr42].

For any decomposition $\omega_n K\mathcal{S}_n = \bigoplus_{p \vdash n} a_p M_p$ into irreducible $\mathcal{S}_n$-modules,

$$L_n(V) = \omega_n T_n(V) \cong \omega_n K\mathcal{S}_n \otimes_{K\mathcal{S}_n} T_n(V) = \bigoplus_{p \vdash n} a_p(M_p \otimes_{K\mathcal{S}_n} T_n(V)).$$

By Schur's fundamental result, $M_p \otimes_{K\mathcal{S}_n} T_n(V)$ is either 0 or an irreducible $\mathsf{GL}(V)$-module, hence the $\mathsf{GL}(V)$-module structure of $L_n(V)$ is completely determined by the multiplicities a_p of the *Lie module*

$$L_n = \omega_n K\mathcal{S}_n$$

of $\mathcal{S}_n$. In his pioneering paper [Kly74], Klyachko also proved the multiplication rules $\omega_n \kappa_n = n\kappa_n$ and $\kappa_n \omega_n = n\omega_n$. Combined with 15.3, these identities imply that

$$L_n = \omega_n K\mathcal{S}_n = \kappa_n K\mathcal{S}_n \cong \iota K\mathcal{S}_n.$$

The isomorphism on the right is left multiplication with $\frac{1}{n}\iota$. It follows that $(\psi_1)^{\mathcal{S}_n}$ is the character of L_n and that

$$L_n(V) = \bigoplus_{p \vdash n} \mathsf{syt}_1^p(M_p \otimes_{K\mathcal{S}_n} T_n(V)),$$

by 15.7. In this spirit, Thrall's problem was finally solved by the work of Klyachko, Kraśkiewicz and Weyman.

Appendix A

Elements of Representation Theory

We start with a proof of

Maschke's Theorem. *If G is a finite group and K is a field of characteristic not dividing the order of G, then KG is semi-simple.*

Proof. Let I be a right ideal of KG and choose a linear projection φ from KG onto I, that is, a surjective linear map $\varphi : KG \to I$ such that $\varphi^2 = \varphi$. Let $\varrho_g : KG \to KG$ denote right multiplication with g, for all $g \in G$, and define a linear map $KG \to KG$ by

$$\tilde{\varphi} := \sum_{g \in G} \varrho_{g^{-1}} \varphi \varrho_g \, .$$

For any $u \in KG$, $h \in G$, we then have

$$(uh)\tilde{\varphi} = u \sum_{g \in G} \varrho_h \varrho_{g^{-1}} \varphi \varrho_g = u \sum_{g \in G} \varrho_{hg^{-1}} \varphi \varrho_g = u \sum_{g \in G} \varrho_{g^{-1}} \varphi \varrho_{gh} = (u\tilde{\varphi})h$$

and

$$u\tilde{\varphi} = \sum_{g \in G} (ug^{-1})\varphi g \in \sum_{g \in G} Ig \subseteq I.$$

Thus $\tilde{\varphi}$ is a homomorphism of KG-modules, mapping KG into I. In particular, the kernel J of $\tilde{\varphi}$ is also a right ideal of KG.

If $u \in I$, then also $u\varrho_{g^{-1}} = ug^{-1} \in I$ and hence $u\varrho_{g^{-1}}\varphi = u\varrho_{g^{-1}}$, for all $g \in G$. This implies $u\tilde{\varphi} = |G|u$ for all $u \in I$.

As a consequence, $1/|G|\tilde{\varphi}$ is a projection of KG-modules from KG onto I, hence $KG = I \oplus J$ and J is a KG-complement of I in KG. □

A.1 Semi-simple Algebras

A short account is now given of the general structure theory of semi-simple algebras. Maschke's Theorem will allow us to apply the results to the group ring KG of any finite group G later on.

Throughout, A is an associative algebra with identity element 1_A. For the sake of clarity, finiteness of dimension and semi-simplicity of A are assumed only where necessary.

Let M, N be A-modules. If N is an A-submodule of M, write $N \leq_A M$. A linear mapping $\varphi : M \to N$ is an *A-homomorphism* if $(m\varphi)a = (ma)\varphi$ for all $m \in M$, $a \in A$. If, in addition, φ is bijective, we say that φ is an *A-isomorphism* and write $M \cong_A N$.

A.1.1 Schur's Lemma. *Let M, N be A-modules, and assume that M is irreducible. If $\varphi : M \to N$ is a nonzero A-homomorphism, then $M\varphi$ is irreducible and $\varphi : M \to M\varphi$ is an A-isomorphism. In particular, the ring*

$$\mathsf{End}_A M := \left\{ \varphi : M \to M \,\middle|\, \varphi \text{ is an } A\text{-homomorphism} \right\}$$

of A-endomorphisms of M is a skew field. If, in addition, K is algebraically closed, then $\mathsf{End}_A M = K\mathsf{id}_M$.

Proof. The kernel $\mathsf{ker}\,\varphi$ is an A-submodule of M. The irreducibility of M implies $\mathsf{ker}\,\varphi = 0$, hence $\varphi : M \to M\varphi$ is bijective. In particular, $M\varphi$ is also irreducible.

If $M = N$ and K is algebraically closed, then there exists an eigenvalue λ of φ in K. For the endomorphism $\varphi - \lambda\mathsf{id}_M \in \mathsf{End}_A M$,

$$0 \neq \mathsf{ker}(\varphi - \lambda\mathsf{id}_M) \leq_A M,$$

and therefore $\varphi = \lambda\mathsf{id}_M$. □

A typical application of Schur's lemma is the following.

A.1.2 Corollary. *Let I be an irreducible (that is, minimal) right ideal of A and $x \in A$, then the right ideal*

$$xI := \{\, xy \,|\, y \in I \,\}$$

of A is either 0 or A-isomorphic to I.

Proof. Apply Schur's lemma to the A-homomorphism $I \to A$, $a \mapsto xa$. □

A.1.3 Proposition. *Let A be semi-simple. For any irreducible A-module M, there exists an irreducible right ideal I of A such that $I \cong_A M$.*

Proof. Let $m \in M$ such that $m \neq 0_M$. Then $m = m1_A \in mA \leq_A M$ and so $mA = M$, as M is irreducible. In other words, the A-module homomorphism

$$\mu : A \to M,\ x \mapsto mx,$$

is surjective. Furthermore, $J := \ker\mu$ is a right ideal of A, that is, an A-submodule of A^R. Now semi-simplicity of A implies that there exists a submodule I of A^R complementing J. As $I \cap J = 0$ and $I + J = A$, it follows that $\mu|_I : I \to M$ is injective and $I\mu = (I + J)\mu = A\mu = M$. Hence $\mu|_I$ is an isomorphism of A-modules and, in particular, I is irreducible. □

Let M be an irreducible A-module, then the *homogeneous component of type M in A* is

$$A_M := \sum\left\{U \leq_A A^R \,\middle|\, U \cong_A M\right\}.$$

Note that, if A is semi-simple, then $A_M \neq 0$, by the preceding proposition.

A.1.4 Proposition. *Let M be an irreducible A-module, then A_M is an ideal of A.*

Proof. By definition, A_M is a right ideal of A.

Let $x \in A$ and choose a right ideal I of A such that $M \cong_A I$, then $xI = 0$ or $xI \cong_A I \cong_A M$, by A.1.2, and therefore $xI \subseteq A_M$ in either case. Hence A_M is also a left ideal of A. □

A.1.5 Theorem. *Let A be semi-simple, and let T be a transversal of the isomorphism classes of irreducible A-modules. Then T is finite, and*

$$A = \bigoplus_{M \in \mathsf{T}} A_M\,.$$

Proof. Semi-simplicity of A allows one to write A as an inner direct sum of a certain set of irreducible right ideals of A. By definition, each of these is contained in a homogeneous component A_M of A, for some $M \in \mathsf{T}$. Hence $A = \sum_M A_M$. Furthermore, by A.1.3, $A_M \neq 0$ for all $M \in \mathsf{T}$.

Let $M \in \mathsf{T}$. Again by semi-simplicity, there exists a right ideal J of A complementing A_M. Let $e \in A_M$ and $f \in J$ such that $1_A = e + f$. Let I

be an irreducible right ideal of A such that $I \not\cong_A M$, then

$$I'I := \sum_{x \in I'} xI = 0$$

for any right ideal I' of A such that $I' \cong_A M$. For, otherwise, $xI \neq 0$ for some $x \in I'$ and thus $I \cong_A xI \subseteq I'$, by A.1.2, contradicting the choice of I. It follows that $A_M I = \sum_{I' \cong_A M} I'I = 0$ and in particular, $eI = 0$. Hence, for any $x \in I$,

$$x = 1_A x = ex + fx = fx \in Jx \subseteq J$$

and therefore $I \subseteq J$. This shows that J contains the sum U of all homogeneous components A_N, $N \in \mathsf{T}\backslash\{M\}$, in A. In fact, it follows that $J = U$, as $J \cap A_M = 0$ and $A = A_M + U$. In particular, $A_M \cap U = A_M \cap J = 0$, which shows that A is the inner direct sum of the family $(A_M)_{M \in \mathsf{T}}$.

Choose $M_1, \ldots, M_k \in \mathsf{T}$ and $e_j \in A_{M_j}$, $1 \leq j \leq k$, such that $1_A = e_1 + \cdots + e_k$. Assume that there exists an A-module $M \in \mathsf{T}\backslash\{M_1, \ldots, M_k\}$, then for all $x \in A_M$,

$$x = x1_A = xe_1 + \cdots + xe_k \in (A_{M_1} + \cdots + A_{M_k}) \cap A_M = 0,$$

hence $A_M = 0$, a contradiction. It follows that $\mathsf{T} = \{M_1, \ldots, M_k\}$ is finite as asserted. □

The *centre of* A, defined by

$$Z(A) := \{ a \in A \,|\, ab = ba \text{ for all } b \in A \},$$

is a commutative subalgebra of A containing 1_A.

A.1.6 Corollary. *Let A be semi-simple, and let T be a transversal of the isomorphism classes of irreducible A-modules. If, for any $M \in \mathsf{T}$, $e_M \in A_M$ is so chosen that*

$$1_A = \sum_{M \in \mathsf{T}} e_M ,$$

then $e_M \in Z(A)$ and

$$e_M e_N = \begin{cases} e_M & \text{if } M = N \\ 0 & \text{otherwise,} \end{cases}$$

for all $M, N \in \mathsf{T}$. Furthermore, e_M is the identity of the subalgebra A_M of A, for all $M \in \mathsf{T}$.

Proof. If $M, N \in \mathsf{T}$ such that $M \neq N$ and $a_M \in A_M$, $a_N \in A_N$, then the product $a_M a_N$ is contained in $A_M \cap A_N$, since A_M and A_N are ideals of A, hence vanishes, by A.1.5.

This implies $e_M e_N = 0$ whenever $M \neq N$ and thus also $e_M = e_M 1_A = \sum_N e_M e_N = e_M e_M$, for all $M \in \mathsf{T}$. More generally, it follows that $a_M e_N = e_N a_M = 0$ whenever $a_M \in A_M$ and $M \neq N$, hence $e_M a_M = 1_A a_M = a_M = a_M 1_A = a_M e_M$ for all $a_M \in A_M$. In particular, $A_M \neq 0$ implies $e_M \neq 0$.

Finally, if a is an arbitrary element of A, then there exist elements $a_N \in A_N$ $(N \in \mathsf{T})$, such that $a = \sum_N a_N$, by A.1.5. It follows that $e_M a = e_M a_M = a_M e_M = a e_M$, hence $e_M \in Z(A)$, for all $M \in \mathsf{T}$. □

An element $e \in A$ is an *idempotent* if $e \neq 0$ and $e^2 = e$. Furthermore, two idempotents $e, f \in A$ are *orthogonal* if $ef = 0_A = fe$. Accordingly, the preceding result states that the elements e_M, $M \in \mathsf{T}$, are mutually orthogonal idempotents in the centre of A. They are the *central primitive idempotents of* A.

A more detailed analysis of the case in which A is of finite dimension over K concludes the first part of this appendix.

A.1.7 Proposition. *Assume that A is finite-dimensional and M is an irreducible A-module, then there exist right ideals $I_1, \dots, I_k$ of A each of which is A-isomorphic to M, such that A_M is the inner direct sum of $(I_1, \dots, I_k)$.*

Proof. By definition, there are right ideals $I_1, \dots, I_n$ of A such that $A_M = \sum_j I_j$ and $I_j \cong_A M$ for all $1 \leq j \leq n$. If the sum of the family $(I_1, \dots, I_n)$ is not direct, there is an index m such that $I_m \cap J \neq 0$, where $J = \sum_{j \neq m} I_j$. Hence $I_m \cap J = I_m$, since I_m is irreducible, and so $I_m \subseteq J$. It follows that $A_M = J$ and after a finite number of steps a family of submodules decomposing A_M directly. □

For any algebra $(B, \cdot, +)$, denote the *opposite algebra* $(B, \bar{\cdot}, +)$ of B by B^{op}, where the product $\bar{\cdot}$ is defined by $a \,\bar{\cdot}\, b := b \cdot a$ for all $a, b \in B$.

A.1.8 Theorem. (Wedderburn) *Let A be semi-simple and finite dimensional, and let T be a transversal of the isomorphism classes of irreducible A-modules, then the algebra A is isomorphic to the direct sum*

$$\bigoplus_{M \in \mathsf{T}} \left((\mathsf{End}_A M)^{k_M \times k_M} \right)^{op}$$

of matrix rings over skew fields, where $k_M := \dim_K A_M / \dim_K M$ denotes the multiplicity of M in A_M, *for all $M \in \mathsf{T}$.*

Proof. By Schur's lemma, $\mathsf{End}_A M$ is a skew field, for all $M \in \mathsf{T}$. Furthermore, A.1.4 and A.1.5 imply $\mathsf{End}_A M = \mathsf{End}_{A_M} M$.

Again by A.1.5, it thus suffices to consider the case where $A = A_M$ for some $M \in \mathsf{T}$, and to show that

$$A \cong \left((\mathsf{End}_A M)^{k\times k}\right)^{op}$$

in this case, where $k = \dim A/\dim M$. Choose irreducible right ideals $I_1, \dots, I_k$ of A according to A.1.7 such that $I_j \cong_A M$ for each j and A is the inner direct sum of the family $(I_1, \dots, I_k)$. (Considering dimensions yields that indeed k direct summands will occur here.)

For any $b \in A$, left multiplication with b, defined by $\lambda_b : A \to A$, $a \mapsto ba$, is an A-endomorphism of the regular A-module A^R. Furthermore,

$$A \longrightarrow \mathsf{End}_A A^R,\ b \longmapsto \lambda_b$$

is an anti-isomorphism of algebras with inverse given by $\varphi \longmapsto 1_A\varphi$. For each $1 \le j \le k$, let $\iota_j : I_j \to M$ be an A-isomorphism and $\pi_j : A \to I_j$ the projection onto I_j (along $\sum_{i\neq j} I_i$). Then for all $\varphi \in \mathsf{End}_A A^R$,

$$\varphi_{ij} := \iota_i^{-1}\,\varphi\,\pi_j\,\iota_j \in \mathsf{End}_A M.$$

Furthermore, as is readily seen, the mapping $\varphi \longmapsto (\varphi_{ij})_{1\le i,j\le k}$ is an isomorphism of algebras $\mathsf{End}_A A^R \to (\mathsf{End}_A M)^{k\times k}$. □

A.2 Finite Group Characters, the Basics

Throughout, G is a finite group and K is a field of characteristic zero. Any G-module M is assumed to be of finite dimension.

A.2.1 Definition and Remark. Let M be a G-module, and let

$$d_M : G \longrightarrow \mathsf{GL}_K(M),\ g \longmapsto (gd_M : m \mapsto mg)$$

be the corresponding representation of G. The trace of the endomorphism gd_M of M is denoted by $\chi_M(g) := \mathsf{tr}(gd_M)$, for all $g \in G$. Then the mapping

$$\chi_M : G \to K,\ g \mapsto \chi_M(g)$$

is the *character of G afforded by M* (or *the character of M*) and

$$\deg \chi_M := \dim_K M$$

its *degree*. The character χ_M is *irreducible* if the module M is irreducible.

A.2.2 Examples.

(i) The character of the regular module $(KG)^R$ is called the *regular character* of G. For any $g \in G$, the matrix of gd_{KG} with respect to the linear basis G of KG is a permutation matrix. As $hg \neq h$ for all $g \in G \setminus \{1_G\}$, $h \in G$, we obtain

$$\chi_{KG}(g) = \begin{cases} |G| & \text{if } g = 1_G\,, \\ 0 & \text{if } g \neq 1_G\,. \end{cases}$$

(ii) The *trivial character* of G, defined by $g \mapsto 1$ for all $g \in G$, is irreducible. The underlying G-module is the *trivial G-module* and is of dimension 1 over K.

The essential structural information about any G-module M is captured by the character χ_M afforded by M, as shall be shown now. To begin with, observe that

$$\dim_K M = \deg \chi_M = \chi_M(1_G),$$

since $1_G d_M = \mathrm{id}_M$.

Now let M' be a second G-module and suppose $\varphi : M \to M'$ is a G-isomorphism, then $(mg)\varphi = (m\varphi)g$, for all $g \in G$ and all $m \in M$, hence $(gd_M)\varphi = \varphi(gd_{M'})$. In particular,

$$\mathrm{tr}(gd_M) = \mathrm{tr}(\varphi^{-1}(gd_M)\varphi) = \mathrm{tr}(gd_{M'}),$$

for all $g \in G$. This shows:

A.2.3 Proposition. *Characters of isomorphic G-modules coincide.*

Simple properties of the trace mapping also lead to the following observations.

A.2.4 Proposition. *For all G-modules M, M_1, M_2, we have:*

(i) *If M is the inner direct sum of M_1 and M_2, then $\chi_M = \chi_{M_1} + \chi_{M_2}$.*

(ii) *$\chi_{M_1} + \chi_{M_2}$ is the character of the G-module $M_1 \oplus M_2$ (outer direct sum).*

Combining A.2.4 with Maschke's Theorem gives:

A.2.5 Corollary. *Any function $\alpha : G \to K$ is a character of G if and only if there exist irreducible characters $\chi_1, \ldots, \chi_m$ of G and nonnegative integer coefficients $a_1, \ldots, a_m$ such that $\alpha = a_1\chi_1 + \cdots + a_m\chi_m$.*

Furthermore, for any $g, k \in G$ and any G-module M, we have

$$\chi_M(k^{-1}gk) = \mathrm{tr}((kd_M)^{-1}(gd_M)(kd_M)) = \mathrm{tr}(gd_M) = \chi_M(g).$$

Hence χ_M is a *class function* of G in the following sense.

A.2.6 Definition and Remark. Two elements $g, h \in G$ are *conjugate* in G if there exists an element $k \in G$ such that $k^{-1}hk = g$. The corresponding equivalence classes in G are the *conjugacy classes* of G. Any function $\alpha : G \to K$ which is constant on the conjugacy classes of G is a *class function* of G. The linear space of all class functions of G (with ordinary pointwise addition and scalar multiplication) is denoted by $\mathcal{C}\ell_K(G)$. For all $\alpha, \beta \in \mathcal{C}\ell_K(G)$, put

$$(\alpha, \beta)_G := \frac{1}{|G|} \sum_{g \in G} \alpha(g)\, \beta(g^{-1}) \in K.$$

The mapping $(\,\cdot\,,\,\cdot\,)_G : \mathcal{C}\ell_K(G) \times \mathcal{C}\ell_K(G) \to K$ is a regular and symmetric bilinear form on $\mathcal{C}\ell_K(G)$.

A.2.7 Example. Let χ be a character of G. Considering the regular G-character χ_{KG} mentioned in A.2.2(i) gives

$$(\chi, \chi_{KG})_G = \frac{1}{|G|} \sum_{g \in G} \chi(g)\, \chi_{KG}(g^{-1}) = \chi(1_G) = \deg \chi\,.$$

We now turn to the simple but powerful notions of induction and restriction of class functions.

A.2.8 Definition and Remarks. Let U be a subgroup of G. Then, on the one hand, $\alpha|_U \in \mathcal{C}\ell_K(U)$, for all $\alpha \in \mathcal{C}\ell_K(G)$. The K-linear mapping

$$|_U : \mathcal{C}\ell_K(G) \to \mathcal{C}\ell_K(U),\ \alpha \mapsto \alpha|_U$$

is called *restriction*.

On the other hand, for each $\alpha \in K^U := \{\beta \,|\, \beta : U \to K\}$, define a mapping $\alpha^G : G \to K$ by

$$\alpha^G(g) := \frac{1}{|U|} \sum_{\substack{x \in G \\ x^{-1}gx \in U}} \alpha(x^{-1}gx),$$

then $\alpha^G \in \mathcal{C}\ell_K(G)$. The K-linear mapping $(\cdot)^G : K^U \to \mathcal{C}\ell_K(G)$, $\alpha \mapsto \alpha^G$, is called *induction*. The class function α^G of G is said to be *induced by* α.

It is easy to show that induction is transitive as follows: For all $\alpha \in K^U$ and all subgroups V of G such that $U \subseteq V$,

$$(\alpha^V)^G = \alpha^G.$$

Induction and restriction come together in the following important result:

A.2.9 Frobenius' Reciprocity Law. *Let U be a subgroup of G, then, for all $\alpha \in \mathcal{C}\ell_K(U)$ and $\beta \in \mathcal{C}\ell_K(G)$,*

$$(\alpha^G, \beta)_G = (\alpha, \beta|_U)_U \,.$$

The straightforward proof is left to the reader.

Restricted and induced characters are again characters. To see this, let U be a subgroup of G, then the linear subspace $KU := \langle U \rangle_K$ of KG is a group algebra of U over K and a subalgebra of KG. So, on the one hand, any KG-module is also a KU-module, by restriction. On the other hand, for any KU-module N, the tensor product $N \otimes_{KU} KG$ of N with the (KU, KG)-bimodule KG is a KG-module with corresponding representation D such that

$$(n \otimes_{KU} a)(gD) = n \otimes_{KU} (ag),$$

for all $n \in N$, $a \in KG$ and $g \in G$.

A.2.10 Theorem. *Let U be a subgroup of G, M be a G-module and N be a U-module, then:*

(i) $\chi_M|_U$ *is the character of the U-module M.*

(ii) $(\chi_N)^G$ *is the character of the G-module $N \otimes_{KU} KG$.*

Proof. (i) holds by definition. Let B be a linear basis of N and R be a transversal of the right cosets of U in G. Then the set

$$T := \{\, b \otimes_{KU} r \mid (b, r) \in B \times R \,\}$$

is a linear basis of $N \otimes_{KU} KG$. Let $g \in G$, $b \in B$ and $r \in R$. Choose $u \in U$ and $r' \in R$ such that $rg = ur'$, then

$$(b \otimes_{KU} r)g = b \otimes_{KU} (rg) = b \otimes_{KU} (ur') = (bu) \otimes_{KU} r' \,.$$

Expanding $(b \otimes_{KU} r)g$ linearly in the elements of T, it follows that $b \otimes_{KU} r$ occurs only if $r = r'$. More precisely, in this case, the coefficient of $b \otimes_{KU} r$

in $(b \otimes_{KU} r)g$ is equal to the coefficient of b in $bu = b(rgr^{-1})$ when written in the basis B of N. Let $d_N : U \to \mathsf{GL}_K(N)$ be the representation of U corresponding to N, then

$$\begin{aligned}
\mathsf{tr}(gD) &= \sum_{\substack{r\in R\\ rg\in Ur}} \mathsf{tr}\Big((rgr^{-1})d_N\Big) \\
&= \sum_{\substack{r\in R\\ rgr^{-1}\in U}} \chi_N(rgr^{-1}) \\
&= \frac{1}{|U|} \sum_{\substack{r\in R\\ rgr^{-1}\in U}} \sum_{u\in U} \chi_N(u(rgr^{-1})u^{-1}) \\
&= \frac{1}{|U|} \sum_{\substack{x\in G\\ xgx^{-1}\in U}} \chi_N(xgx^{-1}) \\
&= (\chi_N)^G(g),
\end{aligned}$$

as χ_N is a class function of U. □

A.3 Orthogonality Relations

A little-known result due to Frobenius [Fro99, Section 5] is used to translate A.1.6 into the orthogonality relations for the irreducible characters of G.

A.3.1 Proposition. *Let V be a finite dimensional vector space over K, and let φ, ψ be linear endomorphisms of V. If $\varphi^2 = \varphi$ and $(V\varphi)\psi \subseteq V\varphi$, then $\mathsf{tr}(\varphi\psi) = \mathsf{tr}(\psi|_{V\varphi})$.*

Proof. Let $B = (b_1, \ldots, b_n)$ be a basis of V such that $B_1 = (b_1, \ldots, b_k)$ is a basis of $V\varphi$. Denote by $A \in K^{k\times k}$ the matrix of $\psi|_{V\varphi}$ corresponding to the basis B_1, and by E_k the identity of $K^{k\times k}$, then the matrices of φ and ψ corresponding to B are respectively

$$\varphi \sim \begin{pmatrix} E_k & 0 \\ 0 & 0 \end{pmatrix} \quad \text{and} \quad \psi \sim \begin{pmatrix} A & 0 \\ * & * \end{pmatrix}.$$

Hence the matrix of $\varphi\psi$ corresponding to B is given by

$$\varphi\psi \sim \begin{pmatrix} A & 0 \\ 0 & 0 \end{pmatrix}.$$

In particular, $\mathrm{tr}(\varphi\psi) = \mathrm{tr}(A) = \mathrm{tr}(\psi|_{V_\varphi})$. □

A.3.2 Frobenius' Lemma. *Let $e = \sum_{x\in G} e_x\, x \in KG$ be an idempotent and $I = eKG$, then for all $g \in G$:*

$$\chi_I(g) = \sum_{h\in G} e_{hg^{-1}h^{-1}}\,.$$

In particular, $\dim_K I = |G|e_{1_G}$.

Proof. Let $g \in G$ and denote by $\varphi : KG \to KG$, $x \mapsto ex$, left multiplication with e, and by $\psi : KG \to KG$, $x \mapsto xg$ right multiplication with g. Proposition A.3.1 may be applied to φ and ψ. Indeed, as $e^2 = e$ and I is a right ideal of KG, we have $(KG)\varphi = eKG = I$, $\varphi^2 = \varphi$ and $I\psi \subseteq I$. Hence

$$\chi_I(g) = \mathrm{tr}(\psi|_I) = \mathrm{tr}(\varphi\psi).$$

But, for all $h \in G$, $h\varphi\psi = ehg = \sum_{x\in G} e_x\, xhg$. As $xhg = h$ if and only if $x = hg^{-1}h^{-1}$, for all $x, h \in G$, it follows that

$$\mathrm{tr}(\varphi\psi) = \sum_{h\in G} e_{hg^{-1}h^{-1}}\,.$$

In case $g = 1_G$, it follows that $|G|e_{1_G} = \chi_I(1_G) = \dim_K I$. □

The set of all sums $\sum_{g\in C} g$, in KG, where C is a conjugacy class of G, is a linear basis of the centre $Z(KG)$ of KG. In particular,

$$\dim_K Z(KG) = h := \#\{\, C \mid C \text{ conjugacy class of } G\,\}.$$

h is the *class number* of G. Furthermore, each element of the centre of KG is constant on conjugacy classes, so that Frobenius' lemma acquires the following simple form for *central* idempotents.

A.3.3 Corollary. *Let $e \in Z(KG)$ be an idempotent and $I = eKG$, then*

$$e = 1/|G| \sum_{g\in G} \chi_I(g^{-1})g.$$

Combined with A.1.6, this gives the

A.3.4 Orthogonality Relations. *For all irreducible G-modules M, N,*

$$(\chi_M, \chi_N)_G = \begin{cases} (\dim_K M)^2 / \dim_K H & \text{if } M \cong_{KG} N, \\ 0 & \text{if } M \not\cong_{KG} N, \end{cases}$$

where H is the homogeneous component of KG of type M (see A.1.4).

Proof. Let M be an irreducible G-module, then the homogeneous component H of KG of type M is an inner direct sum of $k = \dim_K H / \dim_K M$ copies of M, by A.1.7, hence

$$\chi_H = k\chi_M ,$$

by A.2.4. Let e denote the identity of H. Then, by A.1.6, e is a central idempotent in KG and $H = eKG$.

Choose another irreducible G-module N, denote the homogeneous component of KG of type N by L and the identity of L by f. By A.3.3, the coefficient of 1_G in the product ef is

$$\begin{aligned} \frac{1}{|G|^2} \sum_{gh=1_G} \chi_H(g^{-1})\chi_L(h^{-1}) &= \frac{1}{|G|^2} kl \sum_{gh=1_G} \chi_M(g^{-1})\chi_N(h^{-1}) \\ &= \frac{1}{|G|} kl(\chi_M, \chi_N)_G , \end{aligned}$$

where $l = \dim_K L / \dim_K N$. But ef is equal to e or 0 according as $M \cong_{KG} N$ or not, by A.1.6. It follows that $(\chi_M, \chi_N)_G = 0$ if M and N are not isomorphic, whereas in case $M \cong_{KG} N$, we have $\chi_M = \chi_N$, $k = l$ and

$$(\chi_M, \chi_M)_G = k^2 |G| e_{1_G} = k^2 \dim_K H = (\dim_K M)^2 / \dim_K H,$$

which completes the proof. □

The orthogonality relations imply that the set of irreducible G-characters is linearly independent in $\mathcal{C}\ell_K(G)$. Therefore, combining A.2.3, A.2.4(i) and Maschke's Theorem gives:

A.3.5 Theorem. *Let N, N' be G-modules, then $N \cong_{KG} N'$ if and only if $\chi_N = \chi_{N'}$.*

Another consequence is that the number t of mutually non-isomorphic irreducible G-modules is bounded above by the class number h of G, since the dimension of $\mathcal{C}\ell_K(G)$ is equal to h. In general, $t < h$, as the example of the cyclic group of order 3 over the field $\mathbb{Q}$ of rationals shows. However, there is the following clarifying statement.

A.3.6 Theorem. *Let* T *be a transversal of the irreducible* G*-modules, and denote by* h *the class number of* G*, then the following conditions are equivalent:*

(i) $|\mathsf{T}| = h$;

(ii) $\dim_K \mathrm{End}_{KG} M = 1$ *for all* $M \in \mathsf{T}$;

(iii) $\{\chi_M \mid M \in \mathsf{T}\}$ *is an orthonormal basis of* $\mathcal{C}\ell_K(G)$.

The field K is called a *splitting field* of G if one (and hence all) of the conditions (i)-(iii) in A.3.6 hold.

Note that, if K is algebraically closed, then (ii) holds, by Schur's lemma, hence K is a splitting field of G.

Proof of A.3.6. Denote the homogeneous component of KG of type M by H_M, for all $M \in \mathsf{T}$, and set $k_M := \dim_K H_M / \dim_K M$. Then Wedderburn's structure theorem A.1.8 implies

$$k_M \dim_K M = \dim_K H_M = k_M^2 \dim_K \mathrm{End}_{KG} M.$$

for all $M \in \mathsf{T}$. If (iii) holds, then $(\dim_K M)^2 / \dim_K H_M = (\chi_M, \chi_M)_G = 1$ for all $M \in \mathsf{T}$, according to the orthogonality relations A.3.4, hence $k_M = \dim_K M$ and thus $\dim_K \mathrm{End}_{KG} M = 1$ for all $M \in \mathsf{T}$. So (iii) implies (ii). Note that, conversely, (ii) implies $(\chi_M, \chi_M)_G = 1$ for all $M \in \mathsf{T}$.

Assuming (ii), it actually follows that $\mathrm{End}_{KG} M = K\mathrm{id}_M$ for all $M \in \mathsf{T}$, by Schur's lemma. Hence, on the one hand, the centre of the matrix ring $(\mathrm{End}_{KG} M)^{k_M \times k_M}$ over the field $K\mathrm{id}_M$ consists of scalar multiples of the identity matrix only, for all $M \in \mathsf{T}$. On the other hand, the ideal decomposition $KG = \bigoplus_M H_M$ derived in A.1.5 gives the direct decomposition $Z(KG) = \bigoplus_M Z(H_M)$ of the centre of KG. Apply Wedderburn's theorem once more to obtain

$$h = \dim_K Z(KG) = \sum_{M \in \mathsf{T}} \dim_K Z(H_M) = \sum_{M \in \mathsf{T}} 1 = |\mathsf{T}|$$

and hence (i). Furthermore, (iii) follows, since (i) implies that the irreducible characters of G form a basis of $\mathcal{C}\ell_K(G)$, while (ii) implies their orthonormality, as was mentioned already.

It remains to be shown that (i) implies (iii). Assume first that K is algebraically closed. Then (ii) holds, by Schur's Lemma, and thus also (i) and (iii), as we have just seen.

Now consider an arbitrary field K of characteristic 0, and let L be an algebraically closed extension field of K. Then $Z(KG)$ is a subring of $Z(LG)$. Assume that (i) holds and denote by e_M the identity of H_M, for all $M \in \mathsf{T}$. Then $\{ e_M \mid M \in \mathsf{T} \}$ is a K-basis of $Z(KG)$, by (i) and A.1.6. As $\dim_L Z(LG) = h = \dim_K Z(KG)$, A.1.6 also shows that $\{ e_M \mid M \in \mathsf{T} \}$ is an L-basis of $Z(LG)$, namely the unique basis consisting of orthogonal primitive idempotents. Therefore, $\{ \chi_M \mid M \in \mathsf{T} \}$ is the set of irreducible characters of G over L, by A.3.2 and A.1.7. This implies (iii), and the proof is complete. □

A.3.7 Corollary. *Let K be a splitting field of G, then for any $\alpha \in \mathcal{C}\ell_K(G)$,*

$$\alpha = \sum_{\chi} (\alpha, \chi)_G \, \chi ,$$

where the sum is taken over all irreducible characters χ of G.

The short account of the classical representation theory of finite groups ends here.

Appendix B

Solomon's Mackey Formula

Let $n \in \mathbb{N}$. A short proof follows of the Mackey formula

$$\xi^r \xi^q = \sum_{s \models n} m_q^r(s) \xi^s$$

for Young characters corresponding to compositions q, r of n (with certain nonnegative integers $m_q^r(s)$), and of Solomon's noncommutative analogue

$$\Xi^r \Xi^q = \sum_{s \models n} m_q^r(s) \Xi^s$$

in the group algebra of S_n. These identities imply Solomon's theorem 1.1.

The approach presented here is essentially due to Bidigare [Bid97] (see also [Bro00]).

B.1 Definition. An l-tuple $Q = (Q_1, \ldots, Q_l)$ is an *ordered set partition* of $\underline{n}$ if $Q_1, \ldots, Q_l$ are mutually disjoint and nonempty subsets of $\underline{n}$ such that $\underline{n} = Q_1 \cup \cdots \cup Q_l$. Let $q_i = |Q_i|$ for all $i \in \underline{l}$, then

$$\mathsf{type}\, Q := q_1.q_2.\ldots.q_l$$

is a composition of n, the *type* of Q. For example, the set partition P^q of $\underline{n}$ consisting of the successive segments of size $q_1, q_2, \ldots, q_l$ in $\underline{n}$ has type q.

The set of all ordered set partitions of $\underline{n}$ of type q is denoted by Π_q, for all $q \models n$, so that

$$\Pi = \bigcup_{q \models n} \Pi_q$$

is the set of all set partitions of $\underline{n}$.

The natural action of the symmetric group $\mathcal{S}_n$ on $\underline{n}$ extends to an action on Π, via

$$Q\pi = (Q_1\pi, \dots, Q_l\pi)$$

for all $Q = (Q_1, \dots, Q_l) \in \Pi$ and $\pi \in \mathcal{S}_n$. The stabiliser of P^q in $\mathcal{S}_n$ is $\mathcal{S}_q$, the Young subgroup corresponding to q, while the orbit of P^q is Π_q. In particular, the character afforded by the permutation module $K\Pi_q$ is the Young character $\xi^q = (1_{\mathcal{S}_q})^{\mathcal{S}_n}$, that is, $\xi^q(\pi)$ is equal to the number of all set partitions $Q \in \Pi_q$ such that $Q\pi = Q$, for all $\pi \in \mathcal{S}_n$.

B.2 Definition and Remark. Define a product $\wedge$ on Π by

$$(P_1, \dots, P_l) \wedge (Q_1, \dots, Q_k)$$
$$:= (P_1 \cap Q_1, \dots, P_1 \cap Q_k, \ \dots \ , P_l \cap Q_1, \dots, P_l \cap Q_k)^\frown$$

for all $(P_1, \dots, P_l), (Q_1, \dots, Q_k) \in \Pi$, where the hat check on the right hand side indicates that empty sets are deleted. This product turns Π into a semigroup with identity $(\underline{n})$. More importantly, the product $\wedge$ is $\mathcal{S}_n$-*equivariant*, that is,

$$(P \wedge Q)\pi = P\pi \wedge Q\pi,$$

for all $P, Q \in \Pi$, $\pi \in \mathcal{S}_n$.

B.3 Proposition. *The fixed space $\mathcal{B}$ of $\mathcal{S}_n$ in the integral semigroup algebra $\mathbb{Z}\Pi$ is a subalgebra of $\mathbb{Z}\Pi$, with linear basis consisting of the orbit sums*

$$X^q := \sum_{\text{type } Q = q} Q$$

indexed by composition q of n. Furthermore,

$$X^r \wedge X^q = \sum_{s \models n} m_q^r(s) X^s$$

for all $q, r \models n$, where $m_q^r(s) = |\{\, Q \in \Pi_q \mid P^r \wedge Q = P^s \,\}|$.

Proof. If $f, g \in \mathbb{Z}\Pi$ are fixed by $\mathcal{S}_n$, then so is $f \wedge g$, since the product $\wedge$ is $\mathcal{S}_n$-equivariant. Thus $\mathcal{B}$ is a subalgebra of $\mathbb{Z}\Pi$. The orbit sums X^q, $q \models n$, clearly constitute a linear basis of $\mathcal{B}$.

Let $q, r, s \models n$. To describe the structure constants $m_q^r(s)$, it suffices to consider the coefficient of any $S \in \Pi$ of type s in the product $X^r \wedge X^q$. Take $S = P^s$, then $m_q^r(s)$ is equal to the number of pairs $(R, Q) \in \Pi_r \times \Pi_q$

such that $R \wedge Q = P^s$, by definition. However, the latter identity already implies $R = P^r$. □

B.4 Corollary. *Let $\pi \in \mathcal{S}_n$, then the linear span $\mathcal{B}_\pi$ of the elements*

$$X^{q,\pi} := \sum_{\substack{Q \in \Pi_q \\ Q\pi = Q}} Q,$$

indexed by composition q of n, is a subalgebra of $\mathbb{Z}\Pi$. Furthermore,

$$X^{r,\pi} \wedge X^{q,\pi} = \sum_{s \models n} m_q^r(s) X^{s,\pi}$$

for all $q, r \models n$.

Proof. Let $Q, R \in \Pi$, then $R \wedge Q$ is fixed by π if and only if both Q and R are fixed by π. Thus, for $q, r \models n$, the identity

$$X^r \wedge X^q = \sum_{\text{type } R = r} \sum_{\text{type } Q = q} R \wedge Q = \sum_{s \models n} m_q^r(s) X^s$$

stated in B.3 will turn into the identity $X^{r,\pi} \wedge X^{q,\pi} = \sum_{s \models n} m_q^r(s) X^{s,\pi}$ if only those summands are considered which are fixed by π. □

The preceding results allow us to deduce the multiplication rules we are aiming at, as follows.

First, let $\pi \in \mathcal{S}_n$ and consider the sum of the coefficients on both sides of the equation

$$X^{r,\pi} \wedge X^{q,\pi} = \sum_{s \models n} m_q^r(s) X^{s,\pi}.$$

On the left hand side, we get the number of pairs $(R, Q) \in \Pi_r \times \Pi_q$ such that $R\pi = R$ and $Q\pi = Q$, that is, $\xi^r(\pi)\xi^q(\pi)$. Similarly, on the right hand side, the number of summands of $X^{s,\pi}$ is equal to the number of $S \in \Pi_s$ such that $S\pi = S$, hence the coefficient sum is $\sum_{s \models n} m_q^r(s)\xi^s(\pi)$ here. This proves $\xi^r(\pi)\xi^q(\pi) = \sum_{s \models n} m_q^r(s)\xi^s(\pi)$, for all $\pi \in \mathcal{S}_n$, hence the Mackey formula for the Young characters ξ^r, ξ^q.

To prove the noncommutative analogue, B.3 is transferred into the group ring $\mathbb{Z}\mathcal{S}_n$.

The linear space $\mathbb{Z}\Pi_{1^n}$ of $\mathbb{Z}\Pi$ is a two-sided ideal of $\mathbb{Z}\Pi$, by definition of $\wedge$, and an $\mathcal{S}_n$-submodule of $\mathbb{Z}\Pi$. More precisely, the bijection from Π_{1^n}

onto $\mathcal{S}_n$ mapping $(\{a_1\},\ldots,\{a_n\})$ to the permutation $\pi \in \mathcal{S}_n$ with $i\pi = a_i$ for all $i \in \underline{n}$ extends to an isomorphism of $\mathcal{S}_n$-modules

$$\iota : \mathbb{Z}\Pi_{1^n} \to \mathbb{Z}\mathcal{S}_n\,,$$

by linearity. Each element of $\mathcal{B}$ acts on $\mathbb{Z}\Pi_{1^n}$, by left multiplication, and this action commutes with the action of $\mathcal{S}_n$, since $(f\wedge a)\pi = f\pi\wedge a\pi = f\wedge a\pi$ for all $f \in \mathcal{B}$, $a \in \mathbb{Z}\Pi_{1^n}$, $\pi \in \mathcal{S}_n$. In other words, the map

$$\beta_1 : f \mapsto (\varphi : a \mapsto f \wedge a)$$

defines an anti-homomorphism of algebras from $\mathcal{B}$ into $\mathsf{End}_{\mathcal{S}_n}\mathbb{Z}\Pi_{1^n}$. This is isomorphic to $\mathsf{End}_{\mathcal{S}_n}\mathbb{Z}\mathcal{S}_n$, via

$$\beta_2 : \varphi \mapsto \tilde{\varphi} := \iota^{-1}\varphi\iota\,.$$

Finally, there is the usual anti-isomorphism of algebras from $\mathsf{End}_{\mathcal{S}_n}\mathbb{Z}\mathcal{S}_n$ onto $\mathbb{Z}\mathcal{S}_n$, defined by

$$\beta_3 : \tilde{\varphi} \mapsto \mathsf{id}_n\tilde{\varphi}.$$

The composition $\beta = \beta_1\beta_2\beta_3$ of these three maps is a homomorphism of algebras from $\mathcal{B}$ into $\mathbb{Z}\mathcal{S}_n$ such that

$$X^q \mapsto \Big(X^q \wedge (\{1\},\ldots,\{n\})\Big)\iota = \Xi^q$$

for all $q \models n$. Applying β to the multiplication rule for the elements X^q stated in B.3 gives

$$\Xi^r\Xi^q = \sum_{s\models n} m_q^r(s)\Xi^s$$

as desired.

B.5 Remark. Let $q = q_1.\ldots.q_k, r = r_1.\ldots.r_l, s \models n$. As already mentioned in 12.12, the structure constant $m_q^r(s)$ of $\mathcal{D}_n$ can be described combinatorially as the number of certain matrices with nonnegative integer entries. To see this, we use the description of $m_q^r(s)$ given in B.3. Assigning to each $Q \in \Pi_q$ such that $P^r \wedge Q = P^s$ the matrix $M = (m_{ij}) \in \mathcal{M}_q^r$ defined by

$$m_{ij} = |P_i^r \cap Q_j|$$

for all $i \in \underline{l}$, $j \in \underline{k}$, there is a bijection from $\{\, Q \in \Pi_q \mid P^r \wedge Q = P^s \,\}$ onto the set of all $M \in \mathcal{M}_q^r$ such that the word s is obtained by juxtaposing the rows of M from top to bottom and deleting the zeros.

Appendix C

Young Tableaux and Knuth Relations

A proof follows of the Robinson–Schensted correspondence stated in Chapter 8, and its supplement 8.12. Throughout, $n \in {}^{*}$ is fixed.

The approach is inspired by the illustrations of the Robinson–Schensted correspondence for $\mathcal{S}_4$ given in 8.5 and is subdivided into four parts. In the first part, which builds on the theory of frames established in Chapter 6, the noncommutative orthogonality relations 8.10 for the elements Z^p, $p \vdash n$, are derived. The second part follows [BJ99] and contains a combinatorial structure theorem on the Greene cells $\mathcal{G}^p$ in the spirit of 8.5 which is solely based on the definition of plactic and coplactic equivalence. A counting argument due to Leeuwen [Lee96] then allows us to prove the Robinson–Schensted correspondence in the third part. In final part four, a brief study of the so-called Greene invariant leads to a proof of 8.12.

Before starting, recall that σ is a plactic neighbour of π if there exists an index $i \in \underline{n-1}\rfloor$ such that $\sigma = \tau_{n,i}\pi$ and $(i-1)\pi$ or $(i+2)\pi$ is contained in the interval $\langle i\pi, (i+1)\pi \rangle$, for all $\pi, \sigma \in \mathcal{S}_n$ (see 8.1). It will be convenient to write

$$\sigma \;{}_{K}\!\smile\; \pi$$

in this case, and to denote the smallest equivalence on $\mathcal{S}_n$ containing ${}_{K}\!\smile$, the plactic equivalence, by ${}_{K}\!\sim$. Similarly, write

$$\sigma \smile_{K} \pi$$

if σ is a coplactic neighbour of π, that is, if $\sigma^{-1} \;{}_{K}\!\smile\; \pi^{-1}$ or, equivalently, if there exists an index $i \in \underline{n-1}\rfloor$ such that $\sigma = \pi\tau_{n,i}$ and $i-1$ or $i+2$ is contained in $\langle i, i+1 \rangle_\pi$ (see 8.3). The coplactic equivalence is denoted by $\sim_K$, so that $\sigma \sim_K \pi$ if and only if $\sigma^{-1} \;{}_{K}\!\sim\; \pi^{-1}$. The subscript K refers to Knuth.

C.1 Young Tableaux

C.1.1 Theorem. SYT^p *is a coplactic class in* $\mathcal{S}_n$*, for all* $p \vdash n$.

Proof. Lemma 8.7 implies that SYT^p is a union of coplactic classes (without recourse to the Robinson–Schensted correspondence). We use induction on n to show that, conversely, $\pi \sim_K \rho$ for all $\pi, \rho \in \mathsf{SYT}^p$.

This is clear if $n \leq 2$. Let $n \geq 3$ and $\pi, \rho \in \mathsf{SYT}^p$. Set $F := \mathsf{F}(p)$, $\alpha := \iota_F^{-1}\pi$ and $\beta := \iota_F^{-1}\rho$, then both $n\alpha^{-1}$ and $n\beta^{-1}$ are maximal elements of $(F, \leq_F)$, since $n\alpha^{-1} \leq_F y$ implies $n \leq y\alpha$ and thus $y\alpha = n$ for all $y \in F$, and similarly for β. In particular, by 6.11, there exists a partition q such that $F\backslash\{n\alpha^{-1}\} = \mathsf{F}(q)$. Furthermore, $\pi' := \iota_{\mathsf{F}(q)}\alpha \in \mathsf{SYT}^q$. The image line of π' is obtained by removing n from the image line of π. In case $n\alpha^{-1} = n\beta^{-1}$, it follows that

$$\pi' \sim_K \rho' := \iota_{\mathsf{F}(q)}\beta,$$

by induction, hence also $\pi \sim_K \rho$.

Assume that $n\alpha^{-1} \neq n\beta^{-1}$, and let z be the infimum of $n\alpha^{-1}$ and $n\beta^{-1}$ in $(F, \leq_F)$. The set

$$G := \{ w \in F \mid z \leq_F w \}$$

is a partition frame containing z, and so is $G\backslash\{n\alpha^{-1}, n\beta^{-1}\}$. In particular, there is a maximal x element of $G\backslash\{n\alpha^{-1}, n\beta^{-1}\}$ with respect to $\leq_{\mathbb{Z}\times\mathbb{Z}}$. Notice that $n\alpha^{-1} \to x \to n\beta^{-1}$, or $n\beta^{-1} \to x \to n\alpha^{-1}$. Furthermore, x is also a maximal element of $H := F\backslash\{n\alpha^{-1}, n\beta^{-1}\}$ with respect to $\leq_{\mathbb{Z}\times\mathbb{Z}}$. In particular, there is an element $\sigma \in \mathsf{SYT}^H$ such that $(n-2)\delta^{-1} = x$, where $\delta := \iota_H^{-1}\sigma : H \to \underline{n-2}$.

Now define $\gamma : F \to \underline{n}$ by $\gamma|_H := \delta$, $(n-1)\gamma^{-1} := n\beta^{-1}$ and $n\gamma^{-1} := n\alpha^{-1}$, so γ is monotone, by the maximality of $n\beta^{-1} \neq n\alpha^{-1}$. This implies $\nu := \iota_F\gamma \in \mathsf{SYT}^p$. From $n\gamma^{-1} = n\alpha^{-1}$ it follows that

$$\pi \sim_K \nu,$$

as was already shown above. Furthermore, the choice of x implies that $n-2 \in \langle n-1, n\rangle_\nu$ and so, by definition,

$$\nu \smile_K \nu\tau_{n-1} \in \mathsf{SYT}^p.$$

Finally, $n(\gamma\tau_{n-1})^{-1} = n\beta^{-1}$, hence

$$\nu\tau_{n-1} \sim_K \rho$$

as above. Combining all three observations, it follows that $\pi \sim_K \rho$. $\square$

C.1.2 Lemma. *Let $p \vdash n$, then there exists a standard Young tableau $\pi \in \mathsf{SYT}^p$ such that $\pi = \pi^{-1}$.*

Proof. Let $F = \mathsf{F}(p)$, $p = p_1.\dots.p_l$ and $p' = p'_1.\dots.p'_m$. Denote by σ the mapping $F \to F$ reflecting each of the columns F^j, $j \in \underline{m}$, at its centre:

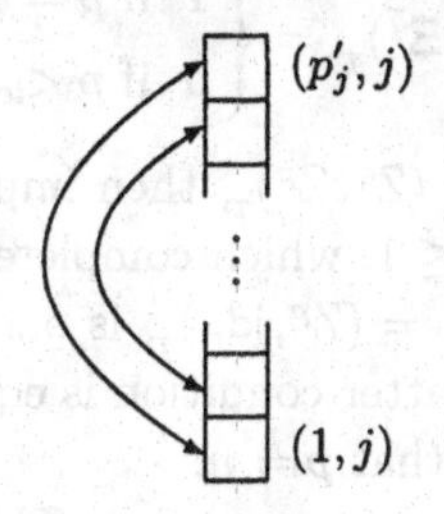

More precisely, define $\sigma : F \to F$ by $(i,j)\sigma := (p'_i + 1 - i, j)$, for all $i \in \underline{l}$, $j \in \underline{p_i}$. As is immediate from the definition, $\sigma^2 = \mathsf{id}_F$. Furthermore, $\sigma : (F, \leq_F) \to (F, \rightarrow_F)$ is monotone. For, if $(i,j), (u,v) \in F$ such that $(i,j) \leq_{\mathbb{Z}\times\mathbb{Z}} (u,v)$, then $i \leq u$ implies $p'_i \geq p'_u$ and hence $p'_i + 1 - i \geq p'_u + 1 - u$, so $j \leq v$ allows one to conclude $(i,j)\sigma = (p'_i + 1 - i, j) \rightarrow (p'_u + 1 - u, v) = (u,v)\sigma$.

As a consequence, $\pi := \iota_F \sigma \iota_F^{-1}$ is contained in $\mathsf{SYT}^F = \mathsf{SYT}^p$, since $\iota_F^{-1}\pi = \sigma\iota_F^{-1}$ is monotone from $(F, \leq_F)$ to $(\underline{n}, \leq)$. Furthermore, $\sigma^2 = \mathsf{id}_F$ implies that $\pi^2 = \mathsf{id}_n$. $\square$

For example, enter the numbers $1, \dots, 9$ in the partition frame $F = \mathsf{F}(4.3.2)$ according to the natural labeling of F to obtain

$$\begin{array}{|c|c|c|c|} \hline 1 & 2 \\ \cline{1-2} 3 & 4 & 5 \\ \cline{1-3} 6 & 7 & 8 & 9 \\ \hline \end{array} \qquad \text{and hence} \qquad \begin{array}{|c|c|c|c|} \hline 6 & 7 \\ \cline{1-2} 3 & 4 & 8 \\ \cline{1-3} 1 & 2 & 5 & 9 \\ \hline \end{array},$$

by reflecting each of the columns at its centre. Reading out the entries again gives $\pi = 6\,7\,3\,4\,8\,1\,2\,5\,9 \in \mathcal{S}_9$. In fact, $\pi \in \mathsf{SYT}^{4.3.2}$ and $\pi^2 = \mathsf{id}_9$. Let it be mentioned that, for an *arbitrary* frame F, the same construction gives an element $\pi \in \mathsf{SYT}^F$ such that $\pi = \pi^{-1}$.

The induction and restriction rules of Chapter 6 now allow us to derive the noncommutative orthogonality relations for the elements Z^p, $p \vdash n$.

C.1.3 Theorem. *Let $n \in \mathbb{N}$ and $p, q \vdash n$, then*

$$(\mathrm{Z}^p, \mathrm{Z}^q)_{\mathcal{P}} = \begin{cases} 1 & \text{if } p = q, \\ 0 & \text{otherwise.} \end{cases}$$

Proof. Let $p = p_1.\dots.p_k$ and $q = q_1.\dots.q_l$ be partitions of n. By C.1.2, $(Z^q, Z^q)_{\mathcal{P}} \geq 1$.

Set $\overline{q} := q_l.\dots.q_1$, then 6.4 implies $\mathsf{SYT}^q \subseteq \mathcal{S}^{\overline{q}}$, hence $(Z^p, Z^q)_{\mathcal{P}} \leq (Z^p, \Xi^{\overline{q}})_{\mathcal{P}}$. An argument similar to the one given in the proof of 10.1 follows, which shows that

$$(Z^p, \Xi^{\overline{q}})_{\mathcal{P}} = \begin{cases} 1 & \text{if } p = q, \\ 0 & \text{if } p <_{\text{lex}} q. \end{cases}$$

The symmetry $(Z^p, Z^q)_{\mathcal{P}} = (Z^q, Z^p)_{\mathcal{P}}$ then implies $(Z^p, Z^q)_{\mathcal{P}} = 0$ if $p \neq q$, and furthermore $(Z^q, Z^q)_{\mathcal{P}} \leq 1$, which completes the proof.

If $q = n$, then $(Z^p, \Xi^{\overline{q}})_{\mathcal{P}} = (Z^p, \mathsf{id}_n)_{\mathcal{P}}$ is one or zero according as $\mathsf{id}_n \in \mathsf{SYT}^p$ or not. By 6.17, the latter condition is equivalent to saying that $\mathsf{F}(p)$ is a horizontal strip, hence that $p = n$.

Assume that $l > 1$ and set $s := q_l.\dots.q_2$. Then, applying 6.15, 5.14, 6.13 and 6.17,

$$(Z^p, \Xi^{\overline{q}})_{\mathcal{P}} = (Z^p, \Xi^s \star \Xi^{q_1})_{\mathcal{P}} = \sum_{r \subseteq p} (Z^r, \Xi^s)_{\mathcal{P}} (Z^{p\backslash r}, \mathsf{id}_{q_1})_{\mathcal{P}} = \sum_{r} (Z^r, \Xi^s)_{\mathcal{P}} ,$$

where the latter sum is taken over all partitions $r \subseteq p$ such that $\mathsf{F}(p\backslash r)$ is a horizontal strip of order q_1. In particular, $p_1 < q_1$ implies $(Z^p, \Xi^{\overline{q}})_{\mathcal{P}} = 0$, since in this case even none of the subsets of $\mathsf{F}(p)$ of order q_1 is a horizontal strip. If $p_1 = q_1$, there exists a unique partition $r \subseteq p$ such that $\mathsf{F}(p\backslash r)$ is a horizontal strip of order q_1, namely $r = p_2.\dots.p_k$. Hence $(Z^p, \Xi^{\overline{q}})_{\mathcal{P}} = (Z^r, \Xi^s)_{\mathcal{P}}$ in this case. Induction finishes the proof. □

C.1.4 Example. In case $n = 4$, the fruits of the results derived on the preceding pages may be summarised and illustrated as follows. The sets

$$\begin{aligned} \mathsf{SYT}^4 &= \{1234\}, \\ \mathsf{SYT}^{3.1} &= \{2134, 3124, 4123\}, \\ \mathsf{SYT}^{2.2} &= \{3412, 2413\}, \\ \mathsf{SYT}^{2.1.1} &= \{3214, 4213, 4312\}, \\ \mathsf{SYT}^{1.1.1.1} &= \{4321\} \end{aligned}$$

are coplactic classes in $\mathcal{S}_4$, by C.1.1. Furthermore, for each $p \vdash 4$, SYT^p and $(\mathsf{SYT}^p)^{-1}$ may be considered as coordinate axes which intersect in the unique element $\pi_p \in \mathsf{SYT}^p \cap (\mathsf{SYT}^p)^{-1}$ obtained by entering the identity in the partition frame $\mathsf{F}(p)$ and then reflecting each of the columns at its centre, by C.1.2 and C.1.3. The element π_p is the origin of the coordinate system indexed by p we are after (see 8.5). For instance, if $p = 2.1.1$, then $\pi_p = 3\,2\,1\,4$. If plactic neighbourhood is illustrated by a vertical bar and coplactic neighbourhood is illustrated by a horizontal bar, each of which is labelled by the respective swap position (as in 8.5), then we get

$$\begin{array}{lllll}
3\,2\,1\,4 & \overset{3}{\text{---}} & 4\,2\,1\,3 & \overset{2}{\text{---}} & 4\,3\,1\,2 \\
\big|\,{\scriptstyle 3} & & & & \\
3\,2\,4\,1 & & & & \\
\big|\,{\scriptstyle 2} & & & & \\
3\,4\,2\,1 & & & &
\end{array}$$

C.2 Carpets

The results that follow allow to complete the illustrative picture coming along with the above example, by means of the local *crochet procedure* C.2.2, which was introduced in [BJ99].

Recall that $\rho_n = n\,(n-1)\,(n-2)\cdots 2\,1$ denotes the order reversing involution in $\mathcal{S}_n$.

C.2.1 Proposition. *Let $\pi, \sigma \in \mathcal{S}_n$, then the following conditions are equivalent:*

(i) *π is a plactic neighbour of σ;*

(ii) *$\rho_n\pi$ is a plactic neighbour of $\rho_n\sigma$;*

(iii) *$\pi\rho_n$ is a plactic neighbour of $\sigma\rho_n$.*

Here, the word plactic may be replaced by the word coplactic.

Proof. By definition, $i\rho_n = n+1-i$ for all $i \in \underline{n}$.

Let $\pi, \sigma \in \mathcal{S}_n$ be plactic neighbours, then there exists an index $i \in \underline{n-1}$ such that $\sigma = \tau_{n,i}\,\pi$ and $(i-1)\pi$ or $(i+2)\pi$ is contained in $\langle i\pi, (i+1)\pi\rangle$.

Clearly, $\sigma\rho_n = \tau_{n,i}(\pi\rho_n)$ and $\rho_n\sigma = \rho_n\tau_{n,i}\pi = \tau_{n,n-i}(\rho_n\pi)$. Concerning left action of ρ_n, the identities $(i-1)\pi = (n-i+2)\rho_n\pi$, $(i+1)\pi = (n-i-1)\rho_n\pi$, $i\pi = (n-i+1)\rho_n\pi$ and $(i+1)\pi = (n-i)\rho_n\pi$ imply that $(n-i-1)\rho_n\pi$ or $(n-i+2)\rho_n\pi$ is contained in $\langle (n-i)\rho_n\pi, (n-i+1)\rho_n\pi\rangle$, hence (ii).

Concerning right action of ρ_n, note that, for arbitrary $a, b \in \underline{n}$, the image under ρ_n of the interval $\langle a, b\rangle$ in $\underline{n}$ is the interval $\langle a\rho_n, b\rho_n\rangle$. Therefore $(i-1)\pi\rho_n$ or $(i+2)\rho_n\pi$ is contained in $\langle i\pi\rho_n, (i+1)\pi\rho_n\rangle$. This implies (iii).

The reverse implications follow from $\rho_n^2 = \mathrm{id}_n$.

Considering the inverse permutations, gives the equivalences for the coplactic neighbourhood. □

C.2.2 Lemma. *Let $\pi \in S_n$ and assume that $i, j \in \underline{n-1}|$ so that*

$$\pi\,\tau_{n,i} \smile_K \pi \;{}_K\!\smile \tau_{n,j}\,\pi,$$

then there exists a permutation $\sigma \in S_n$ such that

$$\pi\,\tau_{n,i} \;{}_K\!\smile \sigma \smile_K \tau_{n,j}\,\pi.$$

The following is a useful illustration of the preceding crochet lemma. Illustrating again plactic and coplactic neighbourhood by a vertical and horizontal bar respectively, the assumption in C.2.2 may be represented by the incomplete stitch

$$\begin{array}{ccc} \pi & \overset{i}{\text{———}} & \pi\,\tau_{n,i} \\ \Big|_{j} & & \\ \tau_{n,j}\,\pi & & \end{array},$$

which, by the claim that follows, may be completed out to

$$\begin{array}{ccc} \pi & \overset{i}{\text{———}} & \pi\,\tau_{n,i} \\ \Big|_{j} & & \Big| \\ \tau_{n,j}\,\pi & \text{———} & \sigma \end{array},$$

by an appropriate crochet procedure, namely by the case-by-case proof that follows.

Proof. Let $\pi = \begin{pmatrix} \cdots & j-1 & j & j+1 & j+2 & \cdots \\ \cdots & a & b & c & d & \cdots \end{pmatrix}$, then

$$a \in \langle b, c\rangle \quad \text{or} \quad d \in \langle b, c\rangle,$$

and

$$i-1 \in \langle i, i+1\rangle_\pi \quad \text{or} \quad i+2 \in \langle i, i+1\rangle_\pi.$$

Without loss of generality, by C.2.1, assume that

$$a \in \langle b, c\rangle \quad \text{and} \quad i-1 \in \langle i, i+1\rangle_\pi$$

using $\rho_n\pi$, $\pi\rho_n$, or $\rho_n\pi\rho_n$ instead of π if necessary.

We consider two cases:

case 1. $|\{i, i+1\} \cap \{a, b, c\}| \leq 1$, then

$$a\tau_{n,i} \in \langle b\tau_{n,i}, c\tau_{n,i} \rangle$$

and so $\pi\,\tau_{n,i} \underset{K}{\smile} \tau_{n,j}\,(\pi\,\tau_{n,i})$. Furthermore, $i-1 \notin \langle i, i+1 \rangle_{\tau_{n,j}\,\pi}$ would imply that $\{b, c\} = \{i-1, i+1\}$ and hence $a = i$, a contradiction. This shows $(\tau_{n,j}\,\pi)\,\tau_{n,i} \smile_K \tau_{n,j}\,\pi$. In other words, $\sigma := \tau_{n,j}\,\pi\,\tau_{n,i}$ completes the stitch.

case 2. $\{i, i+1\} \subseteq \{a, b, c\}$, then $i-1 \in \langle i, i+1 \rangle_\pi$ implies $\{i, i+1\} = \{a, c\}$ and $b = i-1$. Since $a \in \langle b, c \rangle$, it follows that $c = i+1$, $a = i$, hence

$$\pi\,\tau_{n,i} \underset{K}{\smile} \tau_{n,j-1}\,(\pi\,\tau_{n,i}) = (\tau_{n,j}\,\pi)\,\tau_{n,i-1} \smile_K \tau_{n,j}\,\pi.$$

In this case, $\sigma := \tau_{n,j-1}\,\pi\,\tau_{n,i}$ completes the stitch, and we are done. □

C.2.3 Lemma. *If $\pi, \sigma, \sigma' \in \mathcal{S}_n$ such that $\pi \smile_K \sigma$, $\pi \smile_K \sigma'$ and $\sigma \underset{K}{\sim} \sigma'$, then $\sigma = \sigma'$.*

In particular, the permutation σ in C.2.2 is unique.

Proof. Let $\sigma = \pi\,\tau_{n,i}$, $\sigma' = \pi\,\tau_{n,l}$. Since $\sigma \underset{K}{\sim} \sigma'$, there exist $\sigma_1, \ldots, \sigma_k \in \mathcal{S}_n$ such that

$$\sigma = \sigma_1 \underset{K}{\smile} \sigma_2 \underset{K}{\smile} \cdots \underset{K}{\smile} \sigma_k = \sigma'.$$

In particular, there are indices $j_2, \ldots, j_k \in \underline{n-1}$ such that $\sigma' = \tau_{n,j_k} \cdots \tau_{n,j_2}\sigma$, hence

$$\pi\tau_{n,l}\tau_{n,i} = \tau_{n,j_k} \cdots \tau_{n,j_2}\pi.$$

Now assume that $i \neq l$, then there exists an index $m \in \underline{n-1}$ such that m and $m+1$ occur in different orders in the image lines of π and $\pi\,\tau_{n,l}\,\tau_{n,i}$. But, as is immediate from the definition, the orders of m and $m+1$ in the image lines of any two plactic neighbours are the same. Hence m and $m+1$ also occur in different orders in the image lines of $\pi\tau_{n,l}\tau_{n,i}$ and $\tau_{n,j_k} \cdots \tau_{n,j_2}\pi$, a contradiction. □

C.2.4 Example. The coordinate system corresponding to the partition $p = 2.1.1$ of $n = 4$ set off in C.1.4 may be rounded out by four uniquely determined stitches, as follows. Start with the upper left corner, or origin,

$\pi = 3\,2\,1\,4$ of

$$\begin{array}{ccccc}
3214 & \overset{3}{\text{———}} & 4213 & \overset{2}{\text{———}} & 4312 \\
\Big|{\scriptstyle 3} & & & & \\
3241 & & & & \\
\Big|{\scriptstyle 2} & & & & \\
3421 & & & &
\end{array}$$

its plactic neighbour $\tau_{4,3}\pi = 3\,2\,4\,1$ and its coplactic neighbour $\pi\tau_{4,3} = 4\,2\,1\,3$. The unique completion of this stitch is given by $\sigma = 4\,2\,3\,1$:

$$\begin{array}{ccccc}
3214 & \overset{3}{\text{———}} & 4213 & \overset{2}{\text{———}} & 4312 \\
\Big|{\scriptstyle 3} & & \Big|{\scriptstyle 3} & & \\
3241 & \overset{3}{\text{———}} & 4231 & & \\
\Big|{\scriptstyle 2} & & & & \\
3421 & & & &
\end{array}$$

Carry on with the second coplactic neighbour $4\,3\,1\,2$ of $4\,2\,1\,3$, together with its plactic neighbour $\sigma = 4\,2\,3\,1$, to get

$$\begin{array}{ccccc}
3214 & \overset{3}{\text{———}} & 4213 & \overset{2}{\text{———}} & 4312 \\
\Big|{\scriptstyle 3} & & \Big|{\scriptstyle 3} & & \Big|{\scriptstyle 2} \\
3241 & \overset{3}{\text{———}} & 4231 & \overset{1}{\text{———}} & 4132 \\
\Big|{\scriptstyle 2} & & & & \\
3421 & & & &
\end{array}$$

Another two such stitches complete the third row and yield

$$\begin{array}{ccccc}
3214 & \overset{3}{\text{———}} & 4213 & \overset{2}{\text{———}} & 4312 \\
\Big|{\scriptstyle 3} & & \Big|{\scriptstyle 3} & & \Big|{\scriptstyle 2} \\
3241 & \overset{3}{\text{———}} & 4231 & \overset{1}{\text{———}} & 4132 \\
\Big|{\scriptstyle 2} & & \Big|{\scriptstyle 1} & & \Big|{\scriptstyle 1} \\
3421 & \overset{2}{\text{———}} & 2431 & \overset{1}{\text{———}} & 1432
\end{array}$$

as in 8.5.

A formal implementation of the crochet procedure follows.

C.2.5 Corollary. *Let B_0, B_1 be plactic classes in $\mathcal{S}_n$. Assume that there exists an element $\beta_0 \in B_0$, and an element $\beta_1 \in B_1$ such that $\beta_0 \smile_K \beta_1$, then the coplactic neighbourhood induces a one-to-one correspondence between the elements of B_0 and B_1. In other words, for any $\pi_0 \in B_0$, there exists a unique $\pi_1 \in B_1$ such that $\pi_0 \smile_K \pi_1$, and vice versa.*

Here, the words coplactic and plactic may be exchanged.

Proof. Let $\pi_0 \in B_0$, then there exist permutations $\alpha^{(0)}, \ldots, \alpha^{(m)} \in B_0$ such that

$$\beta_0 = \alpha^{(0)} \,{}_K\!\smile\, \alpha^{(1)} \,{}_K\!\smile \cdots {}_K\!\smile\, \alpha^{(m)} = \pi_0 \,.$$

Applying C.2.2 a number of times, yields permutations $\gamma^{(0)}, \ldots, \gamma^{(m)} \in B_1$ such that

$$\beta_1 = \gamma^{(0)} \,{}_K\!\smile\, \gamma^{(1)} \,{}_K\!\smile \cdots {}_K\!\smile\, \gamma^{(m)}$$

and $\alpha^{(i)} \smile_K \gamma^{(i)}$, for all $i \in \underline{m}_{\rfloor}$. In particular, the permutation $\pi_1 := \gamma^{(m)}$ is a coplactic neighbour of π_0 in B_1. Furthermore, π_1 is unique with this property, as is immediate from C.2.3. Thus any $\pi_0 \in B_0$ has a unique coplactic neighbour $\pi_1 \in B_1$. Due to the symmetry in B_0 and B_1 and the symmetry of coplactic neighbourhood, the claim is proved.

By considering inverse permutations, the analogous assertion is obtained for coplactic classes and plactic neighbourhood. □

C.2.6 Definition and Remarks. Consider the smallest equivalence on $\mathcal{S}_n$ refining both ${}_K\!\sim$ and $\sim_K$. Inspired by the illustrations, the corresponding equivalence classes are called *carpets*. This means that two permutations π and σ in $\mathcal{S}_n$ are contained in the same carpet in $\mathcal{S}_n$ if and only if there exist permutations $\pi_0, \ldots, \pi_k \in \mathcal{S}_n$ such that $\pi = \pi_0$, $\sigma = \pi_k$ and, for all $i \in \underline{k}_{\rfloor}$, $\pi_{i-1} \,{}_K\!\smile\, \pi_i$ or $\pi_{i-1} \smile_K \pi_i$.

By definition, any carpet is a union of plactic as well as coplactic classes. In particular, for any partition $p \vdash n$, there exists a unique carpet $\mathcal{T}^p$ in $\mathcal{S}_n$ containing SYT^p, by C.1.1.

C.2.7 Corollary. *Let $\mathcal{T}$ be a carpet in $\mathcal{S}_n$. If A_0, A_1 are coplactic and B_0, B_1 are plactic classes in $\mathcal{T}$, then*

$$|A_0 \cap B_0| = |A_1 \cap B_1|.$$

Proof. We first show that $|A_0 \cap B_0| = |A_0 \cap B_1|$. Assume there exist permutations $\beta_1 \in B_1$, $\beta_0 \in B_0$ such that $\beta_1 \smile_K \beta_0$. Applying C.2.5 yields a bijection $\varphi : B_0 \to B_1$ such that $\beta\varphi = \tilde{\beta}$ if and only if $\beta \smile_K \tilde{\beta}$, for all $\beta \in B_0$, $\tilde{\beta} \in B_1$. By restriction, there is an injective mapping $A_0 \cap B_0 \to A_0 \cap B_1$ and so $|A_0 \cap B_0| \leq |A_0 \cap B_1|$. Symmetry in B_0 and B_1 implies equality. For arbitrary B_1, the result follows by induction. Furthermore, the equality $|A_0 \cap B_1| = |A_1 \cap B_1|$ can be deduced along the same lines once the roles of plactic and coplactic classes are exchanged. □

An immediate consequence of C.2.7 and the noncommutative orthogonality relations is:

C.2.8 Corollary. *Let $p \vdash n$, then for all coplactic classes A and all plactic classes B contained in T^p,*

$$|A \cap B| = 1.$$

Furthermore, the carpet T^p consists of syt^p coplactic classes, and of syt^p plactic classes, each of which contains syt^p elements and a unique element π with $\pi = \pi^{-1}$. In particular,

$$|T^p| = (\mathsf{syt}^p)^2.$$

Proof. The carpet T^p is a union of coplactic classes, one of which is, by definition, $A_0 := \mathsf{SYT}^p$. Furthermore, for the plactic class $B_0 := A_0^{-1}$,

$$|A_0 \cap B_0| = (\mathrm{Z}^p, \mathrm{Z}^p)_{\mathcal{P}} = 1,$$

by C.1.3. This implies $B_0 \subseteq T^p$ and thus $|A \cap B| = 1$ for all coplactic classes A and all plactic classes B in T^p, by C.2.7. In particular, any plactic class in T^p has a unique element in common with A_0, while any coplactic class has a unique element in common with B_0. Hence there are precisely $|A_0| = \mathsf{syt}^p = |B_0|$ plactic as well as coplactic classes in T^p. Any two coplactic and any two plactic classes in T^p have the same cardinality, by C.2.5, and hence the cardinality syt^p of A_0 and B_0. Finally, for any plactic or coplactic class C in T^p, $|C \cap C^{-1}| = 1$ implies that C contains a unique element π with $\pi = \pi^{-1}$. □

Let $p, q \vdash n$, then SYT^p is a coplactic class in T^p, while $(\mathsf{SYT}^q)^{-1}$ is a plactic class in T^q. The remaining part of the noncommutative orthogonality relations C.1.3 implies $|\mathsf{SYT}^p \cap (\mathsf{SYT}^q)^{-1}| = (\mathrm{Z}^p, \mathrm{Z}^q)_{\mathcal{P}} = 0$, hence

$$T^p \cap T^q = \emptyset \qquad (*)$$

whenever $p \neq q$, by C.2.8. In other words, $\mathcal{S}_n$ contains the set $\{\, T^p \mid p \vdash n \,\}$ of mutually disjoint "coordinate systems" in the sense of 8.5. However, the problem remains to show that each permutation $\pi \in \mathcal{S}_n$ is in fact contained in T^p for some $p \vdash n$.

C.3 A Counting Argument

Corollary C.2.8 and (*) above imply

$$\sum_{p \vdash n} (\mathsf{syt}^p)^2 = \sum_{p \vdash n} |T^p| = \Big| \bigcup_{p \vdash n} T^p \Big| \leq n!\,.$$

The difference $n! - \sum_{p \vdash n} (\mathsf{syt}^p)^2$ counts the elements of the carpets $T \notin \{\, T^p \mid p \vdash n \,\}$. An argument due to Leeuwen [Lee96] follows which shows that this difference is zero.

C.3.1 Notation. Let p, q be partitions. Denote by $p \cap q$ the unique partition r in $\mathbb{N}^*$ such that $\mathsf{F}(r) = \mathsf{F}(p) \cap \mathsf{F}(q)$, and by $p \cup q$ the unique partition s in $\mathbb{N}^*$ such that $\mathsf{F}(s) = \mathsf{F}(p) \cup \mathsf{F}(q)$. Furthermore, set

$$p^- := \{\, r \vdash n-1 \mid r \subseteq p \,\} \quad \text{and} \quad p^+ := \{\, s \vdash n+1 \mid p \subseteq s \,\}.$$

C.3.2 Proposition. $\mathsf{syt}^p = \sum_{q \in p^-} \mathsf{syt}^q$, *for all* $p \vdash n$.

Proof. This is a consequence of 6.15, 5.14 and 6.13, since

$$\begin{aligned}
\mathsf{syt}^p &= (\mathrm{Z}^p, \Xi^{1^n})_{\mathcal{P}} \\
&= (\mathrm{Z}^p\!\downarrow, \Xi^{1^{n-1}} \otimes \Xi^1)_{\mathcal{P} \otimes \mathcal{P}} \\
&= \sum_{q \subseteq p} (\mathrm{Z}^q, \Xi^{1^{n-1}})_{\mathcal{P}} (\mathrm{Z}^{\mathsf{F}(p \setminus q)}, \Xi^1)_{\mathcal{P}} \\
&= \sum_{q \in p^-} \mathsf{syt}^q.
\end{aligned}$$

□

C.3.3 Proposition. *For all partitions* p, q*:*

(i) $p \in q^+ \iff q \in p^-$;

(ii) $|p^+| = |p^-| + 1$;

(iii) $|p^+ \cap q^+| = |p^- \cap q^-|$ *whenever* $p \neq q$.

Proof. The first statement holds by the definition of q^+ and p^-. Let $n \in \mathbb{N}$ and $p = p_1.\dots.p_l \vdash n$. Putting $p_{l+1} := 0$, we observe that $|p^+| = \#\{\, j \in \underline{l} \,|\, p_j > p_{j+1}\,\} + 1 = |p^-| + 1$, which proves (ii).

Let $r \in p^+ \cap q^+$, then $q \subseteq r$, $p \subseteq r$ and hence also $p \cup q \subseteq r$. However, $|\mathsf{F}(r)\backslash\mathsf{F}(p)| = 1 = |\mathsf{F}(r)\backslash\mathsf{F}(q)|$ implies that

$$|\mathsf{F}(r)\backslash(\mathsf{F}(p) \cup \mathsf{F}(q))| = |(\mathsf{F}(r)\backslash\mathsf{F}(p)) \cap (\mathsf{F}(r)\backslash\mathsf{F}(q))| \in \{0, 1\}.$$

In fact, $|\mathsf{F}(r)\backslash(\mathsf{F}(p) \cup \mathsf{F}(q))| = 0$ and $\mathsf{F}(r) = \mathsf{F}(p) \cup \mathsf{F}(q)$, since $p \neq q$. This shows that either $p^+ \cap q^+ = \emptyset$ (and $|p^+ \cap q^+| = 0$), or $p^+ \cap q^+ = \{p \cup q\}$ (and $|p^+ \cap q^+| = 1$). Along the same lines, it is seen that either $p^- \cap q^- = \emptyset$ (and $|p^- \cap q^-| = 0$), or $p^- \cap q^- = \{p \cap q\}$ (and $|p^- \cap q^-| = 1$).

It remains to show that $|p^+ \cap q^+| = 1$ if and only if $|p^- \cap q^-| = 1$, but this is readily done. □

C.3.4 Lemma. $(n+1)\mathsf{syt}^p = \sum_{q \in p^+} \mathsf{syt}^q$, *for all* $p \vdash n$.

Proof. The assertion is immediate if $n = 0$. Let $n > 0$, then by induction,

$$\begin{aligned}
(n+1)\mathsf{syt}^p &= \mathsf{syt}^p + n \sum_{r \in p^-} \mathsf{syt}^r \quad \text{, by C.3.2} \\
&= \mathsf{syt}^p + \sum_{r \in p^-} \sum_{s \in r^+} \mathsf{syt}^s \\
&= \mathsf{syt}^p + \sum_{s \vdash n} \sum_{r \in p^- \cap s^-} \mathsf{syt}^s \quad \text{, by C.3.3(i)} \\
&= \mathsf{syt}^p + \sum_{s \vdash n} |p^- \cap s^-| \mathsf{syt}^s \\
&= \mathsf{syt}^p + |p^-||\mathsf{syt}^p| + \sum_{\substack{s \vdash n \\ s \neq p}} |p^+ \cap s^+| \mathsf{syt}^s \quad \text{, by C.3.3(iii)} \\
&= |p^+||\mathsf{syt}^p| + \sum_{\substack{s \vdash n \\ s \neq p}} |p^+ \cap s^+| \mathsf{syt}^s \quad \text{, by C.3.3(ii)} \\
&= \sum_{s \vdash n} \sum_{r \in p^+ \cap s^+} \mathsf{syt}^s \\
&= \sum_{r \in p^+} \sum_{s \in r^-} \mathsf{syt}^s \quad \text{, by C.3.3(i)} \\
&= \sum_{r \in p^+} \mathsf{syt}^r.
\end{aligned}$$

□

C.3.5 Theorem. $n! = \sum_{p \vdash n} (\mathsf{syt}^p)^2$.

Proof. Observe that if $n = 0$, both sides are equal to 1. If $n > 0$, it follows by induction and C.3.2, C.3.3(i), C.3.4 that

$$\begin{aligned}
\sum_{p \vdash n} (\mathsf{syt}^p)^2 &= \sum_{p \vdash n} \sum_{q \in p^-} \mathsf{syt}^p \, \mathsf{syt}^q \\
&= \sum_{q \vdash n-1} \Big(\sum_{p \in q^+} \mathsf{syt}^p \Big) \mathsf{syt}^q \\
&= n \sum_{q \vdash n-1} (\mathsf{syt}^q)^2 \\
&= n\,(n-1)! \\
&= n!\,.
\end{aligned}$$

□

This result implies:

C.3.6 Corollary. *$\mathcal{S}_n$ is the disjoint union of the partition carpets $\mathcal{T}^p$, $p \vdash n$.*

Combined with C.2.8, this allows us to give a

Proof of the Robinson–Schensted correspondence. Let $\pi \in \mathcal{S}_n$, then there exists a partition $p \vdash n$ such that $\pi \in \mathcal{T}^p$, by C.3.6. The plactic class B of π is also contained in $\mathcal{T}^p$, hence intersects the set $\bigcup_{q \vdash n} \mathsf{SYT}^q$ in a unique permutation $P(\pi) \in \mathsf{SYT}^p$, by C.2.8. This defines a mapping

$$P : \mathcal{S}_n \to \bigcup_{q \vdash n} \mathsf{SYT}^q$$

such that $P(\pi) \underset{K}{\sim} \pi$ for all $\pi \in \mathcal{S}_n$, since π and $P(\pi)$ are contained in the same plactic class. In fact, the definition implies that $P(\pi) = P(\sigma)$ if and only if π and σ are contained in the same plactic class, that is, if and only if $\pi \underset{K}{\sim} \sigma$.

Let $Q(\pi) := P(\pi^{-1})$ for all $\pi \in \mathcal{S}_n$. If $\pi, \sigma \in \mathcal{S}_n$ such that $P(\pi) = P(\sigma)$ and $Q(\pi) = Q(\sigma)$, then π and σ are contained in the same plactic class B' and π^{-1} and σ^{-1} are contained in the same plactic class B, say. But B and B' are contained in the same carpet $\mathcal{T}^p$ for some $p \vdash n$, so that $B^{-1} \cap B'$ is a singleton, by C.2.8. This implies $\pi = \sigma$, hence the map $\pi \mapsto (P(\pi), Q(\pi))$ is injective.

Surjectivity now follows from C.3.5, or by definition, since for each pair $(\alpha, \beta) \in \mathsf{SYT}^p \times \mathsf{SYT}^p$ the plactic classes B_α and B_β of α and β respectively are contained in $\mathcal{T}^p$, thus again $B_\alpha \cap (B_\beta)^{-1}$ contains a (unique) permutation π, by C.2.8, and $P(\pi) = \alpha$, $Q(\pi) = P(\pi^{-1}) = \beta$ follows.

The proof of the Robinson–Schensted correspondence is complete. □

Note that, as a consequence, $\mathcal{T}^p$ is the Greene cell $\mathcal{G}^p$ defined in 8.4, for all $p \vdash n$.

C.4 The Greene Invariant

For any permutation $\pi \in \mathcal{S}_n$, Greene [Gre74] discovered a way to determine the partition p of n such that $\pi \in \mathcal{G}^p$, as follows.

C.4.1 Definition and Remark. Let $\pi \in \mathcal{S}_n$ and $k \in \underline{n}$. A subset I of $\underline{n}$ then has *π-level* k if it may be written as a union of subsets $I_1, \ldots, I_k$ such that $\pi|_{I_j}$ is increasing for all $j \in \underline{k}$. Denote the maximum among the cardinalities of all subsets of $\underline{n}$ of π-level k by s_k, and let $l \in \underline{n}$ be minimal such that $s_l = n$. Then the *Greene invariant* of π is defined by

$$\mathsf{g}(\pi) := s_1.(s_2 - s_1).\ldots.(s_l - s_{l-1}).$$

Note that $\mathsf{g}(\pi)$ is a composition of n. It is also important to observe that, in general, there is no set partition $\{I_1, \ldots, I_l\}$ of $\underline{n}$ such that $\pi|_{I_j}$ is increasing for all $j \in \underline{l}$ and $\mathsf{g}(\pi) = |I_1|.\ldots.|I_l|$. For instance, the Greene invariant of the permutation

$$\pi = 2\,4\,7\,9\,5\,1\,3\,6\,8 \in \mathcal{S}_9$$

is $\mathsf{g}(\pi) = 5.3.1$. In fact, $I = \{1, 2, 5, 8, 9\}$ is the *unique* subset of $\underline{9}$ of π-level 1 and cardinality 5, while $I' = \{1, 2, 3, 4, 6, 7, 8, 9\} = \{1, 2, 3, 4\} \cup \{6, 7, 8, 9\}$ is the *unique* subset of $\underline{9}$ of π-level 2 and cardinality 8.

Greene's result is:

C.4.2 Theorem. (Greene, 1974) *Let $\pi \in \mathcal{S}_n$, then $p = \mathsf{g}(\pi)$ is a partition of n and $\pi \in \mathcal{G}^p$.*

This has as consequence Theorem 8.12, as we shall show below. The proof of Greene's theorem is done in three steps and does not differ substantially from the original one.

C.4.3 Proposition. $\mathsf{g}(\pi) = \mathsf{g}(\pi^{-1})$, *for all $\pi \in \mathcal{S}_n$.*

Proof. Let $U \subseteq \underline{n}$, then π is increasing on U if and only if π^{-1} is increasing on $U\pi$. Thus, if $I \subseteq \underline{n}$ and $k \in \underline{n}$, then I has π-level k if and only if $I\pi$ has (π^{-1})-level k. This implies the claim. □

C.4.4 Proposition. *Let $\pi, \sigma \in S_n$ such that σ is a coplactic neighbour of π, then $\mathsf{g}(\sigma) = \mathsf{g}(\pi)$.*

Proof. Since the coplactic neighbourhood is a symmetric relation, it suffices to show that, for each subset I of $\underline{n}$ of σ-level k, there exists a subset J of $\underline{n}$ of π-level k such that $|I| = |J|$.

Suppose $I \subseteq \underline{n}$ has σ-level k. If I also has π-level k, there is nothing to be done.

Assume now that I does not have π-level k. Let $i \in \underline{n-1}$ such that $\pi = \sigma\tau_{n,i}$ and $i-1$ or $i+2$ is contained in $\langle i, i+1\rangle_\sigma$. Put $a := i\sigma^{-1}$ and $b = (i+1)\sigma^{-1}$ and choose mutually disjoint subsets $I_1, \ldots, I_k$ of I such that $\sigma|_{I_j}$ is increasing for all $j \in \underline{k}$ and $I = I_1 \cup \cdots \cup I_k$. We may assume that these subsets are mutually disjoint.

Then a and b are both contained in I_u for some $u \in \underline{k}$. In particular, $a < b$. Let $c \in \langle a, b\rangle$ such that $c\sigma \in \{i-1, i+2\}$.

Consider the case $c\sigma = i-1$. If $c \notin I$, then $J := (I \setminus \{a\}) \cup \{c\}$ has π-level k, as is readily seen. If $c \in I_v$ for some $v \in \underline{k}$, define

$$J_u := (I_v \cap \langle c+1, b\rangle) \cup (I_u \setminus \langle c+1, b\rangle),$$
$$J_v := (I_u \cap \langle c+1, b\rangle) \cup (I_v \setminus \langle c+1, b\rangle)$$

and $J_w := I_w$ for all $w \neq u, v$. The (disjoint) union J of these sets then has π-level k and cardinality $|I|$.

An analogous argument can be given in case $c\sigma = i+2$. □

C.4.5 Proposition. *If $p \vdash n$ and $\sigma \in \mathsf{SYT}^p$, then $\mathsf{g}(\sigma) = p$.*

Proof. Let $n \in \mathbb{N}$ and $p = p_1.\ldots.p_l \vdash n$. Denote the i-th partial sum of p by $s_i := p_1 + \cdots + p_i$, for all $i \in \underline{l}$. Then

$$\pi := \Big((s_{l-1}+1) \cdots s_l\Big)\Big((s_{l-2}+1) \cdots s_{l-1}\Big) \cdots \Big(1 \cdots s_1\Big) \in \mathsf{SYT}^p$$

and $\mathsf{g}(\pi) = p$. The claim follows from C.1.1 and C.4.4. □

Combining C.4.3, C.4.4 and C.4.5 with C.3.6, gives C.4.2.

Greene's theorem brings us in a position to state and prove the following slightly extended version of 8.12.

C.4.6 Theorem. *Let $p \vdash n$. If A is a coplactic class in $\mathcal{G}^p$, then $\rho_n A$ and $A\rho_n$ are coplactic classes in $\mathcal{G}^{p'}$.*

Proof. Let A be a coplactic class in $\mathcal{G}^p$, then $\rho_n A$ and $A\rho_n$ are also coplactic classes in $\mathcal{S}_n$, by C.2.1. In particular, $\rho_n\mathcal{G}^p$ and $\mathcal{G}^p\rho_n$ are Greene cells (or carpets) in $\mathcal{S}_n$. To complete the proof, it suffices to show that $\rho_n\mathcal{G}^p = \mathcal{G}^{p'} = \mathcal{G}^p\rho_n$. In fact, it is enough to show that $\mathrm{g}(\varrho_n\pi) = p'$ for some permutation $\pi \in \mathcal{G}^p$, for this implies $\rho_n\mathcal{G}^p \cap \mathcal{G}^{p'} \neq \emptyset$, hence $\rho_n\mathcal{G}^p = \mathcal{G}^{p'}$, and also $\mathcal{G}^p\rho_n = (\rho_n\mathcal{G}^p)^{-1} = (\mathcal{G}^{p'})^{-1} = \mathcal{G}^{p'}$.

Let $p = p_1.\dots.p_l \vdash n$ and denote the i-th partial sum of p by $s_i := p_1 + \cdots + p_i$, for all $i \in \underline{l}$. Consider the permutation

$$\pi := \Big((s_{l-1}+1).\dots.s_l\Big).\Big((s_{l-2}+1).\dots.s_{l-1}\Big).\dots.\Big(1.\dots.s_1\Big) \in \mathsf{SYT}^p \subseteq \mathcal{G}^p$$

again and assume that I is a subset of $\underline{n}$ of $(\rho_n\pi)$-level k of maximal cardinality. Then $I \cap P_i^p$ has cardinality at most k for all $i \in \underline{l}$, since $(\rho_n\pi)|P_i^p$ is decreasing. It follows that

$$\sum_{i=1}^{k} \mathrm{g}(\rho_n\pi)_i = |I| \leq \sum_{i=1}^{l} \min\{k, p_i\} = \sum_{i=1}^{k} p'_i\,,$$

where we used the observation 12.17 on the partial sums of p'.

Conversely, define $I_j := \{j, s_1 + j, \dots, s_{p'_j - 1} + j\}$ for all $j \in \underline{p_1}$, then $|I_j| = p'_j$ and $(\rho_n\pi)|_{I_j}$ is increasing. This implies

$$\sum_{j=1}^{k} \mathrm{g}(\rho_n\pi)_j \geq \Big|\bigcup_{j=1}^{k} I_j\Big| = \sum_{j=1}^{k} |I_j| = \sum_{j=1}^{k} p'_j$$

and completes the proof. □

Bibliography

[AS] M. Aguiar and F. Sottile. Structure of the Malvenuto-Reutenauer Hopf algebra of permutations. Preprint math.CO/020328228.

[BBHT92] F. Bergeron, N. Bergeron, R. B. Howlett, and D. E. Taylor. A decomposition of the descent algebra of a finite Coxeter group. *J. Algebraic Combinatorics*, 1:23–44, 1992.

[Bid97] T. P. Bidigare. Hyperplane Arrangement Face Algebras and their Associated Markov Chains. Ph.d. thesis, University of Michigan, 1997.

[BJ99] D. Blessenohl and A. Jöllenbeck. Variation über ein Thema von Knuth, Robinson, Schensted und Schützenberger. In *Proc. ALCOMA 1999, Gößweinstein, Germany*, pages 55–66, 1999.

[BL93] D. Blessenohl and H. Laue. Algebraic combinatorics related to the free Lie algebra. In *Publ. IRMA Strasbourg*, Actes 29e Séminaire Lotharingien, pages 1–21, 1993.

[Bro00] K. S. Brown. Semigroups, rings, and Markov chains. *J. Theoret. Probab.*, 13(4):871–938, 2000.

[CR62] C. W. Curtis and I. Reiner. *Representation Theory of Finite Groups and Associative Algebras.* Wiley Interscience, 1962.

[DHT02] G. Duchamp, F. Hivert, and J.-Y. Thibon. Noncommutative symmetric functions VI: free quasi-symmetric functions and related algebras. *Int. J. Algebra Comput.*, 12:671–717, 2002.

[DKK97] G. Duchamp, A. Klyachko, and D. Krob. Noncommutative symmetric functions III: Deformations of Cauchy and convolution algebras. *Discrete Mathematics and Theoretical Computer Science*, I(1):159–216, 1997.

[Fou80] H. O. Foulkes. Eulerian numbers, Newcomb's problem and representations of symmetric groups. *Discrete Math.*, 30:3–49, 1980.

[Fro99] F. G. Frobenius. Über die Darstellung der endlichen Gruppen durch lineare Substitutionen II. *Sitzber. Kgl. Preuß. Akad. Wiss.*, pages 482–500, 1899.

[Ful97] W. Fulton. *Young Tableaux*, volume 35 of *London Mathematical Society, Student Texts*. Cambridge University Press, 1997.

[Gal57] D. Gale. A Theorem on Flows in Networks. *Pacific J. Math.*, 7:1073–1082, 1957.

[Gar89] A. M. Garsia. Combinatorics of the free Lie algebra and the symmetric group. In *Analysis, et cetera*, pages 309–382. P.H. Rabinowitz, E.Zehnder (Eds.), New York, London, 1989.

[Gei77] L. Geissinger. Hopf algebras of symmetric functions and class functions. In *Comb. Represent. Groupe symetr., Actes Table Ronde C.N.R.S. Strasbourg 1976*, volume 579 of *Lecture Notes of Mathematics*, pages 168–181, 1977.

[Ges84] I. M. Gessel. Multipartite P-Partitions and Inner Products of Skew Schur Functions. *Contemp. Math.*, 34, 1984.

[GKL+95] I. M. Gelfand, D. Krob, A. Lascoux, B. Leclerc, V. Retakh, and J.-Y. Thibon. Noncommutative symmetric functions. *Adv. in Math.*, 112(2):218–348, 1995.

[GR89] A. M. Garsia and C. Reutenauer. A decomposition of Solomon's descent algebra. *Adv. in Math.*, 77:189–262, 1989.

[Gre74] C. Greene. An extension of Schensted's theorem. *Adv. in Math.*, 14:254–265, 1974.

[JK81] G. James and A. Kerber. *The representation theory of the symmetric group*. Addison–Wesley, Reading, Massachusetts, 1981.

[Jöl98] A. Jöllenbeck. Nichtkommutative Charaktertheorie der symmetrischen Gruppen. Dissertation, Mathematisches Seminar der Christian-Albrechts-Universität zu Kiel, 1998.

[Jöl99] A. Jöllenbeck. Nichtkommutative Charaktertheorie der symmetrischen Gruppen. *Bayreuth. Math. Schr.*, 56:1–41, 1999.

[JR82] S. A. Joni and G.-C. Rota. Coalgebras and Bialgebras in Combinatorics. *Contemp. Math.*, 6:1–47, 1982.

[JS00] A. Jöllenbeck and M. Schocker. On cyclic characters of symmetric groups. *J. Algebraic Combinatorics*, 12:155–161, 2000.

[Ker91] A. Kerber. *Algebraic Combinatorics via Finite Group Actions*. BI-Wiss.-Verl., Mannheim, Wien, Zürich, 1991.

[KLT97] D. Krob, B. Leclerc, and J.-Y. Thibon. Noncommutative symmetric functions II: Transformations of alphabets. *Int. J. Algebra Comput.*, 7(2):181–264, 1997.

[Kly74] A. A. Klyachko. Lie elements in the tensor algebra. *Siberian Math. J.*, 15:914–920, 1974.

[Knu73] D. E. Knuth. *The Art of Computer Programming III: Sorting and Searching.* Addison-Wesley, 1973.

[KT97] D. Krob and J.-Y. Thibon. Noncommutative symmetric functions IV: Quantum linear groups and Hecke algebras at $q = 0$. *J. Algebraic Combinatorics*, 6(4):339–376, 1997.

[KT99] D. Krob and J.-Y. Thibon. Noncommutative symmetric functions V: a degenerate version of $u_q(gl_n)$. *Int. J. Algebra Comput.*, 9:405–430, 1999.

[KW87] W. Kraskiewicz and J. Weyman. Algebra of invariants and the action of a Coxeter element. Preprint, 1987. Math. Inst. Univ. Copernic, Toruń, Poland.

[KW01] W. Kraskiewicz and J. Weyman. Algebra of invariants and the action of a Coxeter element. *Bayreuth. Math. Schr.*, 63:265–284, 2001.

[Lee96] Marc A. A. Leeuwen. The Robinson-Schensted and Schützenberger algorithms, an elementary approach. *Electronical J. Comb.*, 3, (2):R15, 1996.

[Leh96] B. Lehmann. Defektcharaktere. Diplomarbeit, Mathematisches Seminar der Christian-Albrechts-Universität zu Kiel, 1996.

[LR34] D. E. Littlewood and R. Richardson. Group Characters and Algebra. *Philos. Trans. of the Royal Soc. of Lond., Ser. A*, 233:99–141, 1934.

[LS81] A. Lascoux and M. P. Schützenberger. Le monoïde plaxique. In *Noncommutative Structures in algebra and geometric combinatorics, A. de Luca Ed.*, Quaderni della Ricerca Scientifica del C.N.R., Roma, 1981.

[LST96] B. Leclerc, T. Scharf, and J.-Y. Thibon. Noncommutative cyclic characters of symmetric groups. *J. Combin. Theory Ser. A*, 75(1):55–69, 1996.

[Mac16] P. A. MacMahon. *Combinatory Analysis I, II.* Cambridge University Press, reprint by Chelsea Publishing Company, 1960, 1915/16.

[Mac95] I. G. Macdonald. *Symmetric Functions and Hall Polynomials, 2nd ed.* Clarendon Press, Oxford, 1995.

[Mag40] W. Magnus. Über Gruppen und zugeordnete Liesche Ringe. *J. Reine Angew. Math.*, 182:142–149, 1940.

[Mon93] S. Montgomery. *Hopf algebras and their action on rings*, volume 82 of *Regional Conf. Series in Math.* American Math. Soc., 1993.

[MR95] C. Malvenuto and C. Reutenauer. Duality between Quasi-Symmetric Functions and the Solomon Descent Algebra. *J. Algebra*, 177:967–982, 1995.

[PR95] S. Poirier and C. Reutenauer. Algébres de Hopf de Tableaux. *Ann. Sci. Math. Québec*, 19(1):79–90, 1995.

[Reu93] C. Reutenauer. *Free Lie Algebras*, volume 7 of *London Mathematical Society monographs, new series.* Oxford University Press, 1993.

[Rob38] G. B. Robinson. On the Representations of the Symmetric Group. *Amer. J. Math.*, pages 745–760, 1938.

[RS70] E. Ruch and A. Schönhofer. Theorie der Chiralitätsfunktionen. *Theoret. Chim. Acta*, 19:225–287, 1970.

[Ruc75] E. Ruch. The diagram lattice as structural principle. *Theoret. Chim. Acta*, 38:167–183, 1975.

[Rys57] H.J. Ryser. Combinatorial properties of matrices of zeros and ones. *Canad. J. Math.*, 9:371–377, 1957.

[Sch] M. Schocker. The peak algebra of the symmetric group revisited. to appear in *Adv. in Math.*

[Sch01] I. Schur. Über eine Klasse von Matrizen, die sich einer gegebenen Matrix zuordnen lassen. Dissertation, Berlin, 1901.

[Sch27] I. Schur. Über die rationalen Darstellungen der allgemeinen linearen Gruppe. *Sitzber. Königl. Preuß. Ak. Wiss., Physikal.-Math. Klasse*, pages 58–75, 1927.

[Sch61] C. Schensted. Longest Increasing and Decreasing Subsequences. *Canad. J. Math.*, 13:179–191, 1961.

[Sch63] M. P. Schűtzenberger. Quelques remarques sur une construction de Schensted. *Math. Scand.*, 12:117–128, 1963.

[Sol76] L. Solomon. A Mackey formula in the group ring of a Coxeter group. *J. Algebra*, 41:255–268, 1976.

[Spr74] T. A. Springer. Regular elements of finite reflection groups. *Invent. Math.*, 25:159–198, 1974.

[Sta72] R. P. Stanley. Ordered structures and partitions. *Mem. Amer. Math. Soc.*, 119, 1972.

[Swe69] M. E. Sweedler. *Hopf algebras.* Benjamin, New York, 1969.

[Thr42] R. Thrall. On symmetrized Kronecker powers and the structure of the free Lie ring. *Amer. J. Math.*, 64:371–388, 1942.

[vW98] S. van Willigenburg. A proof of Solomon's rule. *J. Algebra*, 206:693–698, 1998.

[Wit37] E. Witt. Treue Darstellung Liescher Ringe. *J. Reine Angew. Math.*, 177:152–160, 1937.

[You28] A. Young. On Quantitative Substitutional Analysis (third paper). *Proc. London Math. Soc.*, 28:255–292, 1928.

[You34] A. Young. On Quantitative Substitutional Analysis (seventh paper). *Proc. London Math. Soc. (2)*, 36:304–368, 1934.

[Zel81a] A. V. Zelevinsky. A generalisation of the Littlewood–Richardson rule and the Robinson–Schensted–Knuth correspondence. *J. Algebra*, 69:82–94, 1981.

[Zel81b] A. V. Zelevinsky. *Representations of Finite Classical Groups, A Hopf Algebra Approach*, volume 869 of *Lecture Notes in Mathematics*. Springer, 1981.

Index of Symbols

Index